Analysis and Interpretation of
Fire Scene Evidence

José R. Almirall
Kenneth G. Furton
Editors

CRC PRESS

Boca Raton London New York Washington, D.C.

Library of Congress Cataloging-in-Publication Data

Analysis and interpretation of fire scene evidence / edited by José R. Almirall and Kenneth G. Furton.
    p. cm.
Includes bibliographical references and index.
ISBN 0-8493-7885-0
    1. Fire investigation. I. Almirall, José R. II. Furton, Kenneth G.

TH9180.A53 2004
363.37'65--dc22

2003065359

Visit the CRC Press Web site at www.crcpress.com

Preface

The field of forensic science is a diverse, interdisciplinary field that is rapidly expanding in terms of public interest and importance in the administration of justice. A primary reason for the increasing importance of scientific evidence in the courtroom is the increasingly individualizing information being provided by advances in scientific techniques and instrumentation. This book represents the inaugural volume of a new series entitled "Forensic Science Techniques," which will focus on recent developments in the rapidly expanding realm of forensic sciences and emphasize the improvements in the scientific techniques utilized. This book focuses on advances in the analysis and interpretation of fire scene evidence used to help solve arson crimes.

The word "arson" comes from the Latin *ardere*, to burn, and the willful setting of fires has been a recognized crime for thousands of years. The earliest attempt by the Romans to create a code of law was the Laws of the Twelve Tables (c. 455 B.C.). This law detailed *incendium*, the crime of setting any object on fire and resulting in endangering a person's property. Severe punishments were dispensed to those found guilty of this crime, including the sentence of death possibly by burning alive. Of course, the methods of detecting malicious burning of property at that time were rudimentary. Today, the sentences for arson are significantly less severe and the methods for detecting indications of arson are becoming increasingly sophisticated and more selective and sensitive than ever before.

This first book in the series focuses on the scientific advances in the analysis and interpretation of fire scene evidence for the investigation of suspicious fires. It is written to assist those who conduct the chemical analysis and interpretation of physical evidence found at the scene of a fire to determine whether there is a presence of ignitable liquid residues (ILRs). The detection and identification of an ILR at the scene of a fire, in and of itself, does not necessarily result in the conclusion that a crime has been committed. The presence of an unexplained ILR at the scene of a fire often does assist the investigation of an arson. Attorneys and judges involved in criminal and civil judicial proceedings may also find the book a useful reference in preparation for these types of cases.

Practicing forensic chemists and students of forensic chemistry will find Chapter 1 and Chapter 2 useful in better understanding the process that occurs before the laboratory analysis of fire scene evidence can begin. The first two chapters are written from the investigator's point of view to aid chemists and others to better understand basic fire dynamics, ignition, heat transfer, and fire scene investigation techniques. This introduction is important because forensic chemists must be aware of the actions taken in the field. Scientists and investigators interface at some point in the process of analyte detection, collection, packaging, and transport to the laboratory. Therefore, it is essential that investigators and forensic chemists maintain excellent communication and collaboration during the process.

The third chapter, entitled "Detection of Ignitable Liquid Residues in Fire Scenes: Accelerant Detection Canine (ADC) Teams and Other Field Tests," describes the field methods used to identify potential evidence at the scene of a fire suspected as arson. The use of biological detectors (canines) is compared to the use of emerging instrumental field tests. Chapter 4, entitled "Essential Tools for the Analytical Laboratory: Facilities, Equipment, and Standard Operating Procedures," is useful to those who are interested in the initial organization of a new laboratory and setting up standard operating procedures, as well as for revising existing laboratories and procedures. The next chapter contains a detailed description of the analytical methods used in the detection and characterization of ILRs from fire debris, and the sixth chapter, entitled "ASTM Approach to Fire Debris Analysis," details the consensus standards widely used in the discipline. Chapter 7 deals with the interpretation of the data generated from the analyses and includes helpful suggestions for report writing and testimony in these types of cases. The final chapter summarizes new developments in extraction and analysis that can be used to improve the detection of ILRs in fire debris and describes current quality assurance methods in fire debris analysis.

We wish to thank the chapter authors for their expertise and contributions to this volume. We also thank the staff of CRC Press, particularly Becky McEldowney and Julie Spadaro, for their persistence, patience, and encouragement.

We wish to thank the many people who have contributed to our education and success, including our parents, academic mentors, colleagues, and students.

Last but by no means least, we wish to thank our families for their support and encouragement even after countless weekends and late nights spent in front of our computers.

José R. Almirall, Ph.D.
Kenneth G. Furton, Ph.D.
Miami, Florida

The Editors

 **José R. Almirall, Ph.D.,** is an Assistant Professor in the Department of Chemistry and Biochemistry, the Associate Director of the International Forensic Research Institute, and the Director of the Graduate Program in Forensic Science at Florida International University in Miami, Florida. He earned a B.S. in chemistry from Florida International University, an M.S. in chemistry from the University of Miami, and a Ph.D. in pure and applied chemistry from the University of Strathclyde in Glasgow, Scotland. He was a practicing forensic scientist at the Miami-Dade Police Department Crime Laboratory in Miami, Florida, for 12 years prior to his academic appointment in 1998. Dr. Almirall has testified in more than 100 criminal trials as an expert witness in the areas of drugs, trace evidence, and arson evidence analyses. Dr. Almirall has authored or co-authored over 40 publications in the field of analytical chemistry and forensic chemistry and has presented over 140 papers and workshops in the U.S., Europe, Central America, Australia, and Japan. The interests of his research group include the development of analytical methods for the detection and analysis of arson evidence, materials characterizations by a variety of methods, and new applications of mass spectrometry in forensic science.

Kenneth G. Furton, Ph.D., is a Professor in the Department of Chemistry and Biochemistry, Associate Dean of Arts and Sciences, and Director of the International Forensic Research Institute at Florida International University (FIU). He earned a B.S. in forensic science at the University of Central Florida in 1983, a Ph.D. in analytical chemistry at Wayne State University in 1986, and completed postdoctoral studies in nuclear chemistry at the University of Wales, Swansea, in 1988 before becoming a faculty member at FIU. Since then, he has directed the research of scores of undergraduate and graduate students and is the author or co-author of more than 300 publications and presentations. Professor Furton's research program has focused on forensic science and separation science, including the development of novel sample preparation methods prior to chromatographic analysis. Recent work includes studying the chemical basis of detector dog alerts to forensic specimens. He has testified in county and federal court in areas including drug analysis and the use of canines as chemical detectors.

Contributors

José R. Almirall
International Forensic Research Institute and
  Department of Chemistry and Biochemistry
Florida International University
Miami, Florida

Carl Chasteen
State of Florida Fire Marshal's Laboratory
Havana, Florida

Julia A. Dolan
Bureau of Alcohol, Tobacco, Firearms, and
  Explosives
National Laboratory Center
Ammendale, Maryland

Kenneth G. Furton
International Forensic Research Institute and
  Department of Chemistry and Biochemistry
Florida International University
Miami, Florida

Ross J. Harper
International Forensic Research Institute and
  Department of Chemistry and Biochemistry
Florida International University
Miami, Florida

Gregg A. Hine
Bureau of Alcohol, Tobacco, Firearms, and
  Explosives
National Laboratory Center
Ammendale, Maryland

Perry M. Koussiafes
State of Florida Fire Marshal's Laboratory
Havana, Florida

Reta Newman
Pinellas County Forensic Laboratory
Largo, Florida

Jeannette Perr
International Forensic Research Institute and
  Department of Chemistry and Biochemistry
Florida International University
Miami, Florida

David T. Sheppard
Bureau of Alcohol, Tobacco, Firearms, and
  Explosives
National Laboratory Center
Ammendale, Maryland

Contents

Fire Dynamics

1

DAVID T. SHEPPARD

Contents

1.1 What Did the Witness See?

A witness states that he saw the fire while it was still small and before it caused fatalities and destruction. What questions should the investigator ask about the fire itself? What will a fire scientist be able to deduce from the witnesses' answers to the questions?

0-8493-7885-0/04/$0.00+$1.50
© 2004 by CRC Press LLC

People have been interested in fire since the beginning of time, but it is only recently that the tools to evaluate it have been available. Unlike other technologies, there are very few useful relationships that can be expressed in simple algebraic forms. This is because in fires most of the parameters of interest change during the duration of interest. This means that in order to answer useful questions in fire science the equations must be solved in their integral form and in practice this means that a computer must be used.

In the simplest terms, fire can be defined as a rapid chemical reaction where a fuel and oxygen combine to produce heat and light. Of course, in practice nothing is simple.

An understanding of the basic physics of fire phenomena can help the fire investigator to interpret the fire scene. It is important to remember that fires are transient phenomena. Fires grow, shrink, and move, and the fire scene is a record of every phase of the fire. The clues and indicators left after a fire are directly related to how long the fire burned. Fires start small and then grow until the fire size is limited by available fuel or available oxygen.

In a fire situation, the heat from the fire acts as the fundamental mechanism that drives the rest of the fire phenomena. Near the fire, it produces heat, light and products of combustion. The heat from the fire is the primary mechanism that makes smoke move. Through buoyant forces the fire acts as a pump that sucks in air from low regions and, by heating the gases, reduces their density and emits the gases as the products of combustion above the fire. The products of combustion leaving the fire have an initial momentum that acts to create airflow above the fire and throughout structures.

All of these mechanisms can be related to the fire power, also known as the heat release rate. In most scenarios, the dominant factors are all a function of the fire power. The airflows leaving the fire start with a quantifiable amount of momentum introduced by the fire. As the air and smoke flow moves further from the fire, the momentum of the flow is decreased by the solid surfaces that the flow moves past. Therefore, one way to look at smoke flow is that smoke only travels as far as the initial momentum added by the fire can carry it. This is the fundamental reason why the smoke from large fires extends to the farthest recesses of a building and smoke from small fires does not propagate long distances.

The parameters that are related to the fire power are flame height, the rate of flame spread, the ignition of adjacent items, the activation time of detectors, and the dispersion of smoke. There are also many other items of interest that depend on the fire power; for example, flashover in a compartment is a direct result of the fire power and of the dimensions of the room and openings into the room.

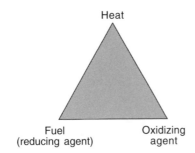

Figure 1.1   Fire triangle.

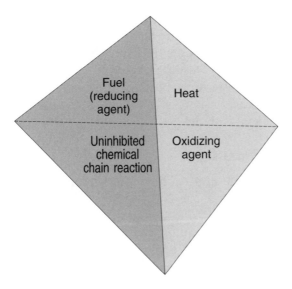

Figure 1.2   Fire tetrahedron.

1.2   Fire

Fire is a rapid oxidation process with the evolution of heat and light in various intensities (NFPA 921). Fire is different from combustion in that fire is usually expected to have an uncontrolled factor to it.

The fire triangle shown in Figure 1.1 and the more recent fire tetrahedron shown in Figure 1.2 are visualization tools that are used to introduce fire science to nontechnical audiences. The basis for these two visualization tools is that for fire to occur and sustain itself there needs to be several available factors. These factors are fuel, oxygen, heat, and chemical chain reactions.

The fire triangle and fire tetrahedron are used first to explain what is needed for a fire to occur and also to explain what can be done to extinguish a fire. Since all of the four factors are required to sustain the fire, the removal of any one factor will effectively extinguish the fire. For example, when there is a natural gas fire, shutting off the fuel supply will stop the fire. If there is a fire in a room, closing the doors can stop the supply of oxygen to the fire and extinguish it. If a fire is doused with water, the water can absorb enough heat to stop the fire. Finally, many chemical fire suppressants impede the fire's chemical chain reaction to stop the fire.

1.2.1   Flame Types

For many fires, gaseous fuel is released from a material due to pyrolysis of a solid fuel or evaporation of a liquid fuel, and the rate of release is a function of the incident heat feedback from the flame and surroundings. The gaseous fuel mixes with an oxidant, usually oxygen from the surrounding air, and when conditions are correct, combustion and visible flame occur. This type of flame is known as a diffusion flame, as opposed to a premixed flame, where the oxidant is integral with the fuel.

1.2.2   Thermochemistry

Thermochemistry is the study of how heat is generated during a chemical reaction. For common fire chemical reactions, the reacting products are a gas-phase fuel and oxygen from the atmosphere. In these cases, the heat generated by the chemical reaction is moderated by the heat absorbed while raising reacting species to their final temperature. Since air is typically the source of oxygen for the fire, and oxygen is only 21% of air (with the majority of the rest being nitrogen), adiabatic flame temperatures from open fires are on the order of 2000°K; cutting torches and other premixed burners that use pure oxygen and not air as the oxidizing source can have much higher flame temperatures on the order of 3000°K.[1]

1.2.3   Heat Release Rate

For most fire scenarios it is much more important to know how quickly the fire is generating heat than how much total heat can be liberated from the fuel. For this reason the essential measure of the fire size is the fire's heat release rate (HRR). Heat release rate is the measure of the power of the fire. Typical units for fires in buildings are kilowatts and megawatts.

The heat release rate from fires is an unsteady phenomenon. For an uncontrolled fire, there is typically a growth phase, a steady burning phase, and a decay phase as the combustible material is fully consumed, as shown in Figure 1.3.

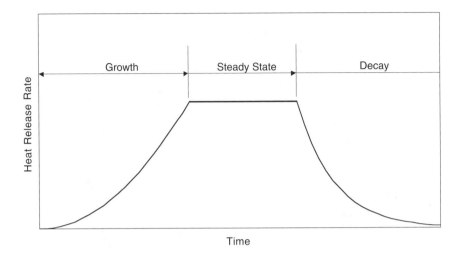

Figure 1.3 Stages of fire growth.

Figure 1.4 Burning candle flame aproximately 0.07 kW.

For realistic fires, the heat release rate can extend over many orders of magnitude. Figure 1.4 and Figure 1.5 show examples of small and large fires, respectively. Figure 1.4 shows the flame of a burning candle; the flame is approximately 10 mm wide × 60 mm tall and has a heat release rate ranging from 0.06 to 0.08 kW. Figure 1.5 shows an example of a large fire created by a heptane spray burner. The fire in Figure 1.5 is approximately 2 m wide × 8 m tall and has a heat release rate of 10,000 kW.

Most simply, the heat release rate is calculated using the relation

$$HRR = \Delta h_c \times \dot{m} \qquad (1.1)$$

Figure 1.5   Heptane spray fire aproximately 10,000 kW.

where Δh_c is the effective heat of combustion in kJ·kg⁻¹ and $\dot{m}$ is the mass loss rate of the fuel in kg·s⁻¹. The relation in Equation 1.1 makes the calculation of heat release rate look straightforward. In practice, the heat of combustion, Δh_c, is not a constant for most common fuels found in fires, but instead varies depending on a number of factors including the geometry of the fuel and the incident heat flux. For this reason, the measurement heat release rate is not a simple process of burning items of known heats of combustion while measuring the weight loss of the sample during burning.

An example of the heat release rate and effective heat of combustion measured for 100 mm × 100 mm × 12.2 mm-thick hardboard is shown in Figure 1.6 and Figure 1.7 using the cone calorimeter (ASTM E1354). The cone calorimeter is a bench-scale test apparatus that exposes the sample surface to a radiant heat flux and measures the heat release rate using the oxygen consumption technique while also measuring the sample mass. The experiment results shown in Figure 1.6 and Figure 1.7 were measured using an external radiant heat flux of 50 kW·m⁻².

In Figure 1.6 the heat release rate remains at zero until ignition occurs at 25 sec. The heat release rate quickly climbs to a local maximum of 2.7 kW at 45 sec. After this initial peak, the heat release rate declines to a steady burning rate of approximately 1.2 kW from 200 to 400 sec. The heat release rate then gradually increases to a peak of 2.7 kW at 670 sec before burning out at approximately 900 sec.

Figure 1.7 shows the effective heat of combustion from the same test shown in Figure 1.6. The effective heat of combustion was calculated by dividing the heat release rate by the mass loss rate measured during the test. It is immediately clear from Figure 1.7 that the heat of combustion is not

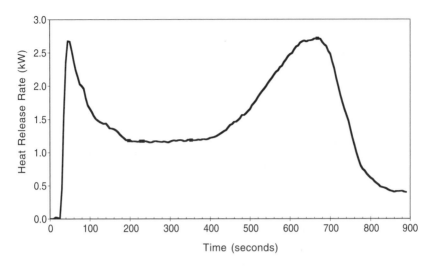

Figure 1.6   Heat release rate of a 100 mm × 100 mm × 12.2 mm-thick sample of hardboard measured in the ASTM E1354 cone calorimeter apparatus at an incident heat flux of 50 kW·m⁻².

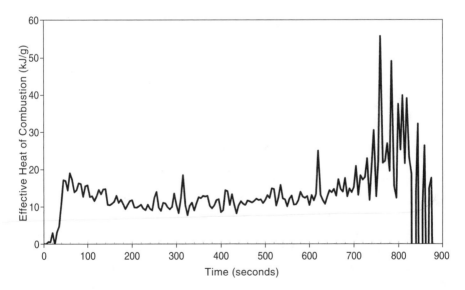

Figure 1.7   Effective heat of combustion rate of a 100 mm × 100 mm × 12.2 mm-thick sample of hardboard measured in the ASTM E1354 cone calorimeter apparatus at an incident heat flux of 50 kW·m⁻².

constant. If the very beginning and end of the test are excluded from the analysis because of measurement errors due to small mass loss rates and only the period of strong burning from 50 to 750 sec is used, the experimental results show that, for this material, the effective heat of combustion varies

Table 1.1   Fire Descriptions

Slow	$Q = \left(0.00293\dfrac{kW}{sec^2}\right)t^2$	(1055 kW in 600 sec)
Medium	$Q = \left(0.01172\dfrac{kW}{sec^2}\right)t^2$	(1055 kW in 300 sec)
Fast	$Q = \left(0.0469\dfrac{kW}{sec^2}\right)t^2$	(1055 kW in 150 sec)
Ultra-fast	$Q = \left(0.1876\dfrac{kW}{sec^2}\right)t^2$	(1055 kW in 75 sec)

by more than 25% during the experiment. For example, from 50 to 100 sec the average effective heat of combustion is 15.7 kJ·g^{-1}, and from 350 to 400 sec it is 11.5 kJ·g^{-1}.

Historically, the growth phase of fires has been generalized in terms of t^2 fires, where the heat release rate grows with the square of time from the start of the fire. Fires have been categorized as four different types, depending on the combustible materials and fire conditions, according to Table 1.1.[2]

The t^2 fire descriptions are empirical generalizations of the heat release rates developed from measurements of real fires. The time for a fire to grow to 1055 kW is also indicated in the table. Because each real fire has a different heat release growth rate, fire protection engineers have found it convenient to design for protection systems to these generalized heat release curves. For example, the engineer may check his fire protection design against medium, fast, and ultrafast fires to assure that the design objectives will be met regardless of the growth rate of the fire.

1.2.3.1   Measuring Heat Release Rate

In early measurements, the heat release rate was calculated from the enthalpy rise method[3] by measuring the temperature rise of the surrounding air (ASTM E1317, ASTM E906). This technique, while relatively simple, was not accurate primarily due to the inability to quantify the fraction of the fire's energy that was radiated away or was lost to the surrounding surfaces.

The current state of the art in heat release rate experiments calculates the heat release rate using measurements of the gas species in the products of combustion from the fire. The ability to calculate the heat release rate using gas species production is a relatively recent development that did not occur until the development of paramagnetic oxygen analyzers, which allow real-time highly accurate measurements of oxygen concentrations. The calculation of the heat release rate of complex items made of many component materials is possible because the amount of oxygen consumed to produce energy in a fire is relatively constant at approximately 13.1 MJ/kg of oxygen

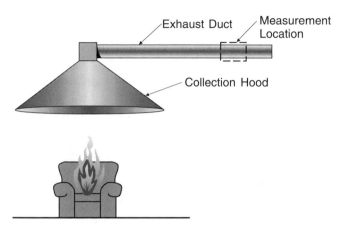

Figure 1.8 Calorimeter.

consumed. The devices used to measure heat release rate using this technique are called oxygen consumption calorimeters because their primary function is to calculate heat release rate by means of accurately measuring the amount of oxygen that has been consumed from the air during the combustion process.

Oxygen consumption calorimeters used for measuring heat release rate share the same basic design for bench-scale through large-scale experiments, as shown in Figure 1.8. The basic design includes a collection hood positioned above the burning item to collect all products of combustion. The collection hood is connected to an exhaust duct with a powered exhaust blower. Within the exhaust duct, measurements are made of the mass flow rate of the gases in the duct and the concentrations of the primary gas constituents (e.g., oxygen, carbon dioxide, and carbon monoxide). In most calorimeters, measurements are also made of the obscuration of a light source due to smoke.

1.2.3.2 Predicting Heat Release Rate

It would be useful to be able to predict heat release rates of real fires from bench-scale test results. In reality, this a difficult calculation because of the many scenario-dependent factors that affect the burning rate. Two of the most important factors affecting burning rate are the thickness of the materials and the incident heat flux. The thickness of the material affects how quickly it can be heated to its ignition temperature and, as a result, dictates the material's pyrolysis rate. The incident heat flux comes from the flames from the burning item and also from the surroundings. Because of the many factors affecting the heat release rate in large-scale burning, the scaling of heat release rate from bench scale to large scale is not practical for most applications at this time.

1.3 Enclosure Fires

Fires in rooms and other enclosures behave differently from fires in the open. The heat from the fire is captured in the enclosure and produces feedback with the fire. This is analogous to pouring water into a bucket with holes where the bucket represents the enclosure and the water represents the heat from the fire. When water is added slowly to the bucket, it has time to flow out the holes and does not collect in the bucket. As the rate of flow increases, the water level in the bucket increases because it cannot flow out of the holes quickly enough to keep the bucket from filling. The weakness of this analogy is that as heat collects in a room, there is heat feedback that causes the fire to grow more quickly and, as a result, increases the heat input into the compartment.

Consider phases of fire growth in a simple residential room. While the fire is small it is called "fuel limited." This means that the size of the fire is restricted by the rate at which it can heat fuel to its ignition temperature. In this fuel-limited state, as the fire grows it is able to heat more fuel close to the fire and the rate of fire growth is primarily restricted by the type of fuel and the geometry of the fuel source. If the fire is located in an enclosure such as a room, the heat and products of combustion from the fire will collect at the top of the room and will start to radiate heat downward to all areas of the room. At some point, the fire can grow to a size where the fire requires more oxygen than is coming in through the openings in the room, and at this point the fire is considered "ventilation limited."

Figure 1.9 shows a schematic of a fire in a compartment. The figure shows a fire on a bed with a plume rising above the flames. The plume transfers the hot gases into the hot gas layer. Once the hot gas layer descends below the height of openings such as doors and windows, the hot gases flow out of the compartment.

There are three stages of compartment fires:

1. *Open fire:* When a fire is small relative to the size of the enclosure, the fire behaves as though the enclosure does not exist. In this situation, there is unrestricted airflow to the fire and the size of the fire is limited only by the availability of fuel. In this case the fire is called fuel limited, which means that the fire growth rate is restricted only by the availability of fuel. As the fire grows and a hot upper layer develops, heat feedback to the fire starts, which causes the fire to burn more quickly than it would in the open. The fire is still fuel limited because there is still enough available air to support the combustion.
2. *Flashover:* There is a transition between when the fire is burning in individual locations and when all of the flammable items in the compartment are burning. This transition is called flashover. Flashover is

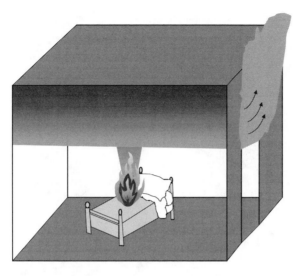

Figure 1.9   Two-layer compartment fire.

indicated by several events including rollover, flaming out of the openings, and off-gassing of combustible items throughout the compartment. Rollover occurs when unburned combustible gases in the upper layer ignite and flames appear to roll across the ceiling. Flaming-out of the openings is caused by insufficient oxygen in the room for combustion. Off-gassing of combustible items in the lower layer is caused by a large increase in the radiant flux.

3. *Fully developed fire:* In a fully developed fire, there is burning of all combustible items in the compartment. In such a case the fire is fuel rich and ventilation limited, meaning that the size of the fire in the compartment is regulated by the amount of oxygen that can enter through the openings in the compartment. The vitiated atmospheres in ventilation-limited fires cause a change in the combustion chemistry, resulting in a dramatic increase in toxic gas production, including carbon monoxide.

Figure 1.10 shows a kitchen fire in its initial open stage. The fire was initiated in a roll of paper towels mounted below the cabinets and allowed to grow up the side of the cabinets near the stove. This fire was ignited approximately 2 min before this picture and the room flashed over approximately 3 min after. This fire is in its initial open stage as evidenced by the lack of a smoke layer.

Figure 1.11 shows a couch fire in a living room. When the picture was taken, the flame was about 0.6 m high and 0.3 m wide. There is a clearly

Figure 1.10 (see color insert following page 54)   Open fire in kitchen cabinets. Courtesy of Special Agent/CFI Michael R. Marquardt, Federal Bureau of Alcohol, Tobacco, Firearms, and Explosives.

defined plume above the fire that extends up to the smoke layer located approximately 1.5 m above the floor. At the time the picture was taken, significant radiant heat was starting to be felt from the upper layer.

The most commonly used techniques for calculating the environment in compartment fires prior to flashover is the use of computer zone models. In a zone model, the environment in compartments is divided into two layers with a hot upper layer above a relatively cool lower layer. The fire plume is used as the primary mechanism for transferring heat and mass from the lower layer to the upper layer. There are several computer zone models in widespread use in the fire protection field. The most widely used zone model in the U.S. is CFAST (Consolidated Model of Fire Growth and Smoke Transport). CFAST is available free of charge from the National Institute of Standards' (NIST) Web site, and training is available from many sources.

Figure 1.12 shows a fully developed fire in a bathroom. The picture was taken from outside the building through an open window. Two firefighters shown in the picture are in a room adjacent to the bathroom. The flames in the bathroom extend from floor to ceiling and out of the door about 2.4 m along the ceiling of the adjacent room. The photographer reported that the radiant flux was painful where the picture was taken.

Figure 1.11 (see color insert following page 54) Open fire in a living room showing the fire plume and hot smoke layer. Courtesy of Special Agent/CFI Michael R. Marquardt, Federal Bureau of Alcohol, Tobacco, Firearms, and Explosives.

1.4   Ignition

Solid and liquid materials do not burn in their initial phase and must first be converted to the gas phase before ignition of the flammable vapors will occur. For liquids this is accomplished by vaporizing the fuel, and for solids the primary mechanism is pyrolysis.

1.4.1   Gaseous Ignition

In the presence of a spark or a flame, gases will ignite if the gas concentration is between the lower flammability limit (LFL) and upper flammability limit (UFL). This flammability range between the upper and lower flammability limits varies widely for various materials. If the gaseous concentration is below the lower flammability limit the mixture is known as fuel lean; when the gaseous concentration is above the upper flammability limit the mixture is known as fuel rich. It is possible in fire scenarios for the flammable gas concentration to build up above the upper flammability limit in compartments. This is a very dangerous situation because if any air is suddenly

Figure 1.12 (see color insert following page 54)　Fully developed fire in a bathroom.

introduced into the compartment, which can occur when a firefighter opens a door or breaks a window, the new air introduced will reduce the concentration into the flammability range and thus cause a fireball or gas-phase explosion.

There are several methods available for measuring the flammability limits of gas mixtures. One of the more commonly used methods was developed at the Bureau of Mines.[4] The apparatus consists of a 1.5-m vertical tube, which is 0.05 m in diameter with a sealed top and bottom. The test procedure consists of uncovering the bottom of the tube and applying a spark or small pilot flame. The mixture is deemed to be flammable if the flame front progresses halfway up the tube. Other common test methods include the ASTM E681 method that uses a spark igniter in a sealed container.

When an external pilot source is not available, the gas can also ignite if the mixture is at or above its autoignition temperature. For many gas mixtures, the minimum autoignition temperature at atmospheric pressure is above 300°C, but for some gas mixtures (e.g., carbon disulfide with a minimum autoignition temperature of 90°C) the autoignition temperature can be much lower.

1.4.2 Liquid Ignition

For a liquid to ignite, it must first evaporate and form a flammable mixture. In the presence of a pilot, such as a spark or small flame, ignition will occur where the evaporated liquid fuel has achieved its lower flammability limit. The location where ignition occurs depends on the ambient airflow and the rate of evaporation. In stagnant conditions with little airflow, the concentration of the evaporated vapors will increase. In an environment with small airflow, the flammable vapors will not increase to the lower flammability limit for many fuels. Sustained flaming depends on whether the evaporation rate is sufficient to maintain a gas concentration above the lower flammability limit. For most situations, the major factor affecting the evaporation rate is the radiant feedback from the flame to the liquid fuel. The minimum temperature at which a flammable mixture will form immediately above a liquid is known as the flashpoint or piloted ignition temperature. When no pilot is present the liquid must be heated to its autoignition temperature. The flashpoint is lower than the boiling point for a liquid and the autoignition temperature is above the boiling point. Therefore, the liquid must be fully converted to a gas before autoignition will occur. A practical example of how this impacts a real fire-ignition scenario is gasoline, which, in the presence of a pilot, will evaporate and provide a flammable mixture at normal room temperatures, but when no pilot is present, the vapor must come into contact with an object with a temperature greater than its autoignition temperature of 440°C before it will ignite.

There are many standard tests for measuring the piloted ignition temperature of liquids, depending on the type of liquid, the flashpoint temperature, and the apparatus used. Most of these test methods fall into two categories: closed cup and open cup (e.g., Tag closed tester [ASTM D56] and Cleveland tester [ASTM D92], respectively). In the closed-cup methods the liquid is held in a container with a loosely fitted cover; in the open-cup methods, liquid is held in a container with no cover. In both types of methods, liquid is uniformly heated and the ignition source is typically a spark. A schematic of these two test types is shown in Figure 1.13.

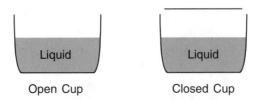

Figure 1.13 Open-cup and closed-cup test methods for measuring liquid flash-point.

1.4.3 Ignition of Solids

Ignition of solid fuels from an external heating source is a complicated process. Solid fuels do not ignite in their original form and must first be decomposed into gaseous products in a process known as pyrolysis. For most common materials, the pyrolysis process is not reversible but instead is path dependent because of the development of char. Once the flammable gaseous products achieve the required concentration, they can be ignited either by a pilot or through autoignition.

The ignition temperature of solids is not typically a constant and instead depends on the method of heating. For practical purposes, ignition temperatures for materials are often measured experimentally, but the user must be aware that the ignition temperature thus measured is dependent on the test method.

One way to measure the autoignition temperature and the piloted ignition temperature of solids is by using the Stetchkin furnace (ASTM D1929) shown in Figure 1.14. A specimen is placed in a 40-mm specimen pan within an adjustable temperature furnace. A temperature rise measured by thermocouples placed near the specimen is used to measure the autoignition temperature. Alternatively, a pilot flame mounted above the furnace is used to determine the flash ignition temperature. The test procedure consists of placing the test specimen in the preheated furnace and observing if ignition occurs. The procedure is repeated until the autoignition or flash ignition temperature is bracketed to an acceptable level of accuracy.

1.4.4 Flame Height

There are few quantifiable measurements that can be derived from witness statements. One of the exceptions is flame height. This information is important because the flame height can provide insight into the heat release rate of the fire.

The momentum of the burning fuel and buoyancy forces govern the flame height. The size of a burning surface can be characterized by a dimension, D, that is characteristic of the area of the burning surface. The velocity

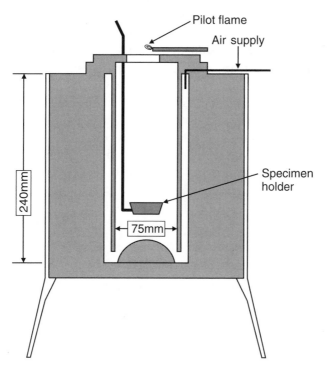

Figure 1.14   Cross-section view of Stetchkin furnace.

at which products are leaving this surface can be characterized by a velocity, U. For pyrolysis reactions, the velocity, U, can be related to the heat release rate based on the heat of combustion of the burning material.[5]

The nondimensional Froude number is typically used when considering applications in which the momentum and buoyancy in a plume are of similar magnitudes. For fire applications, the Froude number is typically defined as shown in Equation 1.2, where C is a constant based on the fuel properties, the gravitational constant, and the fuel geometry (pool of burning liquid, jet flames, etc.).

$$\text{Fr} = U^2 / gD \propto Q^2 / D^5 = C\left(Q/D^{5/2}\right)^2 \tag{1.2}$$

For fires the nondimensional parameter $\dot{Q}^*$ is often used to quantify the Froude number:[6]

$$\dot{Q}^* = \frac{\dot{Q}}{\rho_\infty C_p T_\infty D^2 \sqrt{gD}} \tag{1.3}$$

where $\dot{Q}$ is the total heat release rate of the fire (kW), ρ_{∞} is the ambient air density (kg·m^{-3}), T_{∞} is the ambient air temperature (°C), C_p is the constant pressure specific heat of the air (kJ·kg^{-1}·s^{-1}), g is the acceleration of gravity (m·s^{-2}), and D is the diameter of the fire source (m). The Froude number is often raised to the $\dot{Q}^{*2/5}$ power to linearize the experimental flame height correlations.

Visible flames appear to be fairly steady near the base and become more intermittent in the upper part of the fire. For analysis purposes, researchers define the mean flame height as the location where visible flame is present at least 50% of the time. Flame height correlations compiled by McCaffrey[6] show that $\dot{Q}^{*2/5}$ can range from approximately 0.5 for pool fires to 1000 for jet flames.

For fires in the buoyancy regime, Heskestad has compiled a correlation of mean flame heights which covers the entire range of $\dot{Q}^*$ except for the momentum regime

$$\frac{L}{D} = -1.02 + 3.7\dot{Q}^{*2/5} \tag{1.4}$$

where L is the mean flame height (m) and D is the diameter of the fire source (m). For noncircular fire sources, D can also be the effective diameter such that $\pi D^2/4$ = area of the fire source. At normal atmospheric conditions, Equation 1.4 simplifies to

$$L = -1.02D + A\dot{Q}^{2/5} \tag{1.5}$$

For many fuels, A ranges from 0.226 to 0.240, and for calculation a value of 0.235 is typically used. For some common fuels such as acetylene and gasoline, A can deviate significantly from the given range. For these cases and for a more in-depth discussion of flame heights, Heskestad has provided an excellent reference.

1.5   Fire Plume

A buoyant stream called a fire plume rises above a localized fire area, as shown in Figure 1.15. The fire plume is usually turbulent except in very small fires.[7]

There are many empirical relations for calculating plume temperatures and velocities.[7] For axisymmetric plumes, the plume correlations are based on an analysis by Morton.[8] In this analysis, Morton made the following assumptions:

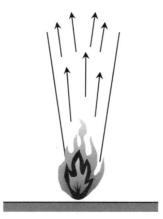

Figure 1.15   Fire plume shown above a fire.

- There is a point source of buoyancy.
- Variations in density in the field of motion are small compared to the ambient density.
- The air entrainment into the plume is proportional to the local vertical plume velocity.
- The profiles of the vertical velocity and the buoyancy force are of similar form at all heights and are axisymmetric.

Upon applying these assumptions, the mean motion in the plume is governed by the conservation equations for mass, momentum, and energy, as shown in Equation 1.6 through Equation 1.8. In these equations, b is the radius of the plume, g is the acceleration of gravity, u is the mean plume velocity, ρ is the mean density of the plume, ρ_∞ refers to the density of the air outside the plume, α is a proportionality constant for the entrainment into the plume, and z is the height.

The continuity equation, Equation 1.6, shows that the increase in the mass flow in the plume as a function of height is equal to the entrainment of gases into the plume.

Continuity: $$\frac{\partial}{\partial z}\left(b^2 u\right) = 2\alpha b u \qquad (1.6)$$

The momentum equation, Equation 1.7, shows that the change in momentum of the plume is balanced by the buoyant force of the gases in the plume.

Momentum: $$\frac{\partial}{\partial z}\left(b^2 u^2\right) = b^2 g \frac{\rho_\infty - \rho}{\rho} \qquad (1.7)$$

The energy equation, Equation 1.8, shows that energy enters only at the source of the plume, so there is no change in energy in the plume.

Energy:
$$\frac{\partial}{\partial z}\left(b^2 u g \frac{\rho_\infty - \rho}{\rho}\right) = 0 \tag{1.8}$$

After making an ideal gas assumption, the convective heat release rate of the fire can be used to relate the temperature rise of the plume to the decrease in density. This leads to the following proportionalities based on Equation 1.6 to Equation 1.8:

$$b \propto z \tag{1.9}$$

$$u \propto Q_c^{1/3} z^{-1/3}$$

$$\Delta T \propto Q_c^{2/3} z^{-5/3}$$

These relations show that the plume radius, b, is proportional to the height, z. The velocity and the temperature rise in the plume are both functions of only the convective heat release rate, Q_c, and the height, z.

Researchers have found these relationships to be accurate above the mean flame height of the fire.[5] The proportionality constants found by researchers for the plume radius range from 0.15 to 0.18. The proportionality constants for the centerline velocity range from 0.8 to 1.2 $(m \cdot s^{-1}(m \cdot kW^{-1})^{1/3})$. The proportionality constants for the centerline temperature range from 21 to 30.5 $(m \cdot °C(m \cdot kW^{-1})^{2/3})$.

To compensate for the fact that real fires have finite areas and are not point sources, as the Morton theory requires, researchers have assumed that there is a virtual point source located either above or below the fire. This virtual point source is referred to as the virtual origin (Figure 1.16). Each researcher has developed a virtual origin equation that best fits his data. Gupta tabulated five equations for the virtual origin.[7] Although the equation that each researcher developed was different from the others, all of the equations are functions only of the heat release rate, Q, and the diameter of the fire, D. The exponent of the heat release rate in these equations ranged from 2/5 to 2.

Heskestad's virtual origin equation is presented in Equation 1.10 because of its simplicity and its central location among the other correlations.[5]

$$z_o = -1.02D + \left(0.083\frac{1}{kW^{2/5}}\right)Q^{2/5} \tag{1.10}$$

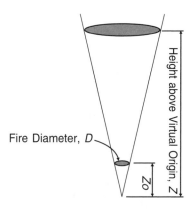

Figure 1.16   Virtual origin.

It is important to note that for small fire sizes the virtual origin is below the fire location, and as the fire grows, the virtual origin moves above the fire location.

The *SFPE Fire Protection Engineering Handbook* suggests empirically derived equations that incorporate the virtual origin used for determining plume centerline temperature and centerline velocity:

$$u_{cl} = 3.4 \left[\frac{g}{C_p \rho_\infty T_\infty} \right]^{1/3} Q_c^{1/3} (z - z_o)^{-1/3} \tag{1.11}$$

$$\Delta T_{cl} = 9.1 \left[\frac{T_\infty}{g C_p^2 \rho_\infty^2} \right]^{1/3} Q_c^{2/3} (z - z_o)^{-5/3} \tag{1.12}$$

Here the *cl* subscript denotes a value at the plume centerline and ∞ denotes a value for the entrained gas. Other values take on their traditional meaning.

For locations not on the plume centerline, the following equations are suggested:

$$\Delta T = \Delta T_{cl} e^{-\left(R / \sigma_{\Delta T} \right)^2} \tag{1.13}$$

$$u = u_{cl} e^{-\left(R / \sigma_u \right)^2} \tag{1.14}$$

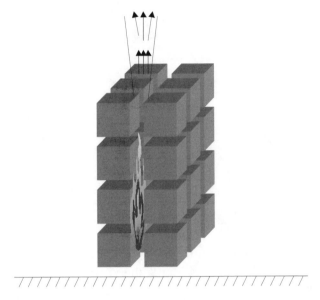

Figure 1.17   Fire plume in rack storage.

where $^\Delta T$ and u are local values at radius R from the centerline of the plume. $\sigma_{\Delta T}$ and σ_u are measures of plume width corresponding to the radii where the local values are e^{-1} multiplied by the centerline values.

Equation 1.12 can be written as a proportionality:

$$^\Delta T \propto Q_c^{2/3}\left(z-z_o\right)^{-5/3} \tag{1.15}$$

A special case, which is of specific interest in sprinkler applications, is the behavior of fire plumes above the rack storage used in warehouses and wholesale/retail stores, as shown in Figure 1.17. These plumes differ from the axisymmetric plumes discussed earlier because the entrainment is limited while the plume is within the rack storage array.

An initial comparison between axisymmetric plumes and rack storage plumes indicates that rack storage plumes should be narrower and hotter and have higher velocities. This should occur because an axisymmetric plume is able to entrain relatively cool air from all directions for its entire height. In rack storage the entrained air can enter the plume only at gaps between stored commodities. Limiting the entrainment limits the cooling in the rack storage plume, making it hotter. Reducing the entrainment rate also causes the plume mass to increase more slowly. The velocities will be larger in the rack storage plume because there is less mass in the plume. Since the only driving force for the velocity is at the bottom of the plume, the plume with slower mass increase will decelerate more slowly. The result is a "chimney effect" in gaps between stored commodities.

Researchers[9,10] have found that the plumes above the stored commodity follow the same functional relationships as axisymmetric plumes, although the constants of proportionality are changed. The equations for the plume temperature and velocity are shown in the following equation:

$$u_{cl} = \left(4.25 \frac{m^{4/3}}{kW^{1/3}s}\right)Q_c^{1/3}(z - z_o)^{-1/3} \qquad (1.16)$$

$$\Delta T_{cl} = \left(11 \frac{{}^\circ C\, m^{5/3}}{kW^{2/3}}\right)Q_c^{2/3}(z - z_o)^{-5/3}$$

The effect of the lack of entrainment in the rack storage configuration is clearly seen in the equations for the virtual origin. The *SFPE Handbook* presents two equations for the virtual origin correction to be used when the fire plume is confined in the flue space between the racks:

Two-tier storage: $z_o = -1.6 + \left(0.094 \dfrac{m}{kW^{2/5}}\right)Q_c^{2/5}$

Three- and four-tier storage: $z_o = -2.4 + \left(0.95 \dfrac{m}{kW^{2/5}}\right)Q_c^{2/5}$ (1.17)

These equations are much different from the virtual origin equations presented for axisymmetric plumes because the virtual origin correction is calculated from the top of the stored commodity rather than from the base of the fire (see Figure 1.18). This is interesting because the virtual origin correction has the effect of moving the heat source higher for even relatively small fires. For example, an infinitely small fire located at the floor in a 5-m-high rack storage would place the virtual origin 3.4 m above the floor.

1.6 Ceiling Jet

When the fire plume reaches a horizontal obstacle such as a ceiling, the vertical buoyancy-driven flow becomes a momentum-driven horizontal flow. In locations where the ceiling is large and unobstructed, this flow will be

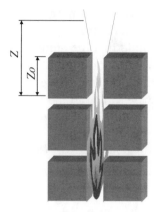

Figure 1.18   Virtual origin for rack storage.

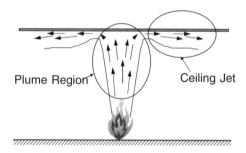

Figure 1.19   Fire plume and ceiling jet.

axisymmetric. This horizontal momentum-driven flow is known as the ceiling jet, shown in Figure 1.19.

An example of the temperatures in the ceiling jet in the presence of a growing fire is shown in the color figure insert following page 54.[11] These temperatures were measured with five thermocouples positioned 4, 8, 12, 24, 36 in. below a 25-ft-high ceiling. The thermocouples were located at a 35-ft radial distance from the fire. In this case it is clear that the highest temperatures are located within 12 in. of the ceiling and that temperatures below this point are significantly lower.

There has been substantial research in this area because of the importance of the ceiling jet in transporting smoke and heat to the fire sprinklers and smoke detectors. Most of the experimental research has been conducted with large, flat, unobstructed ceilings. There has also been some experimental work conducted for ceilings with roof trusses hung below them.

Beyler has compiled empirical correlations of ceiling jet temperatures and velocities from a number of different researchers.[12] Two of these correlations by

Heskestad and Delichatsios[13] are provided in Equation 1.18 through Equation 1.20.

$$Q_0^* = Q \big/ \left(\rho_\infty C_p T_\infty g^{1/2} H^{5/2} \right) \tag{1.18}$$

$$\Delta T_0^* = \Delta T / T_\infty \big/ \left(Q_0^* \right)^{2/3} = \left[0.188 + 0.313 r / H \right]^{-4/3} \tag{1.19}$$

$$U_0^* = 0.68 \left(\Delta T_0^* \right)^{1/2} \left(r / H \right)^{-0.63} \tag{1.20}$$

where Q_0^* and ΔT_0^* are intermediate calculation variables. Q is the fire's heat release rate, ρ_∞ is the density of the ambient air, C_p is the specific heat of the ambient air, T_∞ is the temperature of the ambient air, g is the acceleration due to gravity, H is the height of the ceiling, r is the radial distance from the centerline of the fire, ΔT is the maximum temperature rise in the ceiling jet, and U_0^* is the maximum velocity in the ceiling jet.

Because of the transient nature of the heat losses from the ceiling jet to the ceiling, simple correlations for the temperatures and the velocities in the ceiling jet such as the ones provided in Equation 1.19 and Equation 1.20 can only provide estimates and should not be considered definitive.

1.7 Heat Transfer

Heat from fires is transferred to its surroundings by convection and radiation. The buoyant plume transports heated gases from the fire to the surroundings above the fire. The remaining energy from the fires is transported to the surroundings by radiation.

1.7.1 Radiation

An important mode of heat transfer in fires is radiation.[14] Thermal radiation is the electromagnetic radiation emitted by a body as a result of its temperature.[15] In fires, radiation heat transfer occurs from the flame to the surrounding surfaces as well as between the surfaces themselves. At the beginning of a fire event, the air between surfaces is essentially transparent to the thermal radiation. As smoke begins to fill a space, constituents of the smoke (e.g., water vapor, carbon dioxide, soot particles, etc.) become significant absorbers and emitters of thermal radiation, and the space between the surfaces becomes an important participating medium in the thermal heat transfer.

The total amount of thermal energy emitted by an ideal radiator is defined by the Stefan–Boltzmann law as

$$E_b = \sigma T^4 \tag{1.21}$$

where E_b is in $W \cdot m^{-2}$; σ is the Stefan–Boltzmann constant, which has a value of 5.669×10^{-8} $W \cdot m^{-2} \cdot K^{-4}$; and T is the temperature in degrees Kelvin. The subscript b denotes that the radiation is from a black body.

When calculating the amount of thermal energy that is transmitted from one surface to another, the radiation shape factor (also known as the configuration factor, the view factor, and the angle factor) is used to define the faction of energy that leaves one surface and reaches another. In practice, the radiation shape factor is used as follows:

$$q_{1-2} = A_1 F_{1-2} \left(E_1 - E_2 \right) \tag{1.22}$$

where q_{1-2} is the heat transfer between surfaces 1 and 2, A_1 is the area of surface 1, F_{1-2} is the shape factor between surfaces 1 and 2, and (E_1-E_2) is the difference in the thermal energy emitted by the two surfaces. In Equation 1.21, note that if the temperatures of the two surfaces are the same, the heat transfer between the two bodies is zero.

For black bodies, the major hurdle to calculating radiant heat transfer using Equation 1.22 is determining the appropriate shape factor, which often requires integration for complicated shapes. Unfortunately, most surfaces in real scenarios are not black bodies, and this makes the problem much more difficult to solve because not all of the energy striking a surface is absorbed. Some of the energy striking the surface will be reflected to other surfaces and some of the energy will be lost to the system.

1.7.2　Conduction

Conduction is the heat transfer that occurs within a material. In fire scenarios, conductive heat transfer is typically categorized as either thermally thin or thermally thick.

Thermally thin materials are those materials for which the rate of heat transfer within the object is substantially faster than the rate at which heat transfer changes at the surface such that temperature is essentially uniform within the object. This condition occurs for thin materials or for materials with extremely high thermal conductivities. Examples of thermally thin materials are fabrics, single sheets of paper, and sheet metal. Figure 1.20 shows a cross section of a thermally thin material. On the side of the material

$$(\delta \rho \, C_p)\frac{\partial T}{\partial x} = \dot{q}''_r + \dot{q}''_c - \dot{q}''_e$$

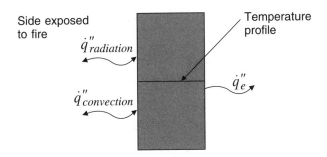

Figure 1.20 Cross section of a thermally thin material displaying a uniform temperature within.

exposed to the fire, the dominant modes of heat transfer are radiation, $\dot{q}''_{radiation}$, and convection, $\dot{q}''_{convection}$. On the unexposed side of the material the heat transfer, $\dot{q}''_e$, will depend on the backing material.

The temperature of thermally thin materials can be described by the one-dimensional heat transfer equation (Equation 1.23). Here the temperature, T, is calculated by integrating over the thickness, x, of the material. A lumped thermal property parameter, $\delta \rho C_p$, representing the product of the thickness, density, and specific heat, respectively, is used in this calculation. The $\delta \rho C_p$ used in the calculation must be measured using a transient temperature test method because static temperature test methods will not produce acceptable results.

$$\left(\delta \rho \, C_p\right)\frac{\partial T}{\partial x} = \dot{q}''_r + \dot{q}''_c - \dot{q}''_e \qquad (1.23)$$

Equation 1.23 can be solved for the time, t_{ig}, for a thermally thin material to achieve its ignition temperature, T_{ig}, as shown in Equation 1.24.

$$t_{ig} = \left(\delta \rho \, C_p\right)\frac{\left(T_{ig} - T_\infty\right)}{\dot{q}''} \qquad (1.24)$$

where T_∞ is the initial material temperature and $\dot{q}''$ is the total incident heat flux on the surface.

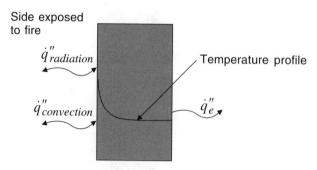

Figure 1.21 Cross section of a thermally thick material showing the temperature gradient within.

Most materials more than 1 mm thick behave as thermally thick in fire scenarios. In thermally thick materials there is a temperature gradient within the material, as shown by the curved line in Figure 1.21. The term "thermally thick" represents that the material is sufficiently thick that heat transfer from the unexposed surface does not significantly effect heat transfer into the exposed surface.

Heat transfer within a thermally thick material with constant thermal properties is governed by the transient heat transfer equation, which for a one-dimensional case is

$$\rho C_p \frac{\partial T}{\partial t} = k \frac{\partial^2 T}{\partial x^2} \tag{1.25}$$

where the new variable k is the thermal conductivity of the material.

When Equation 1.25 is solved for the time, t_{ig}, for the exposed surface to achieve a given ignition temperature, T_{ig},

$$t_{ig} = \frac{\pi}{4} k \rho C_p \frac{\left(T_{ig} - T_\infty\right)}{\dot{q}''^2} \tag{1.26}$$

Although most materials are treated as thermally thick for ignition calculations, the long-term temperature profile approaches a linear relationship as time progresses, as shown in Figure 1.22.

When the temperature profile becomes linear, the heat transfer is considered to be in a steady-state condition and the temperature within the material is defined by Fourier's law

$$\dot{q}'' = -k \frac{\partial T}{\partial x} \tag{1.27}$$

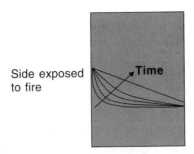

Figure 1.22 Temperature profile within a thermally thick material as time progresses.

1.7.3 Convection

The two fundamental modes of heat transfer are conduction and radiation. Convection is conductive heat transfer through a fluid. For practical considerations, convection is usually treated as a third type of heat transfer because of the difficulty of calculating the conductive heat transfer through a moving fluid. The equation used to calculate the convective heat transfer from a fluid to a surface is

$$\dot{q}'' = h\left(T_\infty - T_s\right) \qquad (.28)$$

where h is the convective heat transfer coefficient, T_∞ is the temperature of the fluid, and T_s is the temperature of the surface.

For calculation purposes the convective heat transfer coefficient is typically an empirically based number that is calculated from a dimensionless Nusselt number and the characteristics of the heat transfer problem. Nusselt numbers for a variety of geometric configurations are provided in most undergraduate heat transfer texts.

1.8 Smoke and Visibility

The smoke properties of primary interest in fires are visibility and obscuration. The smoke particle size distribution along with the amount produced determines the properties of the smoke.[16] The most widely measured smoke property is the light extinction coefficient. The basis of light extinction measurements is Bouguer's law, which defines that the intensity, I, of radiation of wavelength, λ, exponentially decays from an initial intensity, of I_{λ_0} as a function of the optical density, k, and the distance, D.

$$I_\lambda = I_{\lambda_0} e^{-kD} \tag{1.29}$$

or alternatively

$$k = \frac{1}{D} \ln\left(I_{\lambda_0}/I_\lambda\right) \tag{1.30}$$

The optical density, k, has been found to be proportional to the mass concentration of smoke for a specific type of smoke. This finding is important to researchers and experimentalists because it allows the optical density to be treated like a mass concentration for calculations and for experimental measurements. It is also important to realize that the optical density is inversely proportional to the distance, D. For example, when the optical density doubles, the distance at which objects are visible is halved.

Experimentally, the optical density is measured using a light source and a photocell separated by a volume with smoke. The smoke measurements are either taken in a closed chamber (ASTM E662, ISO 5659) or as a dynamic measurement in a calorimeter exhaust duct, as shown in Figure 1.8.

Visibility in a smoke-filled environment is affected by several factors including the visual acuity of the person, the contrast of the object being viewed vs. its background, and the optical density. Visibility experiments have historically been conducted using test subjects viewing a variety of objects in different lighting conditions. From these experiments the maximum distance, S, at which an object will be visible can be calculated using an empirical visibility constant, V, and the optical density as

$$S = V/k \tag{1.31}$$

For example, Mulholland[16] suggests that for light-emitting and light-reflecting signs empirical visibility constants of $V = 8$ and $V = 3$, respectively, can be used.

1.9   Recommended References

There are several references that should be included in every fire scientist's library. *An Introduction to Fire Dynamics* by Dougal Drysdale[17] was the first college textbook on fire dynamics. This textbook, which is in its second edition, provides the scientific background for the development of fire safety engineering. *Principles of Fire Behavior* by James Quintiere[18] uses a quantitative approach to present the scientific principles behind fire behavior. Many of the equations in this textbook have been presented in algebraic rather than integral form, and there are many worked examples of practical engineering problems. The primary reference, which no library should be without, is *The SFPE Handbook of Fire Protection Engineering*.[19] This reference, which is in it third edition, is a comprehensive guide to fire protection engineering. An expert in the subject has written each chapter, and the latest research has been incorporated into the text.

Notes

1. Drysdale, D.D., Thermochemistry, in *The SFPE Handbook of Fire Protection Engineering*, 3rd ed., Sec. 1, Chap. 5, National Fire Protection Association, Quincy, MA, 2002.

2. Evans, D.D., Ceiling Jet Flows, in *The SFPE Handbook of Fire Protection Engineering*, 1st ed., Sec. 1, Chap. 9, National Fire Protection Association, Quincy, MA, 1988.

3. Janssens, M., Calorimetry, in *The SFPE Handbook of Fire Protection Engineering*, 3rd ed., Sec. 3, Chap. 2, National Fire Protection Association, Quincy, MA, 2002.

4. Zabetakis, M.G., Flammability characteristics of combustible gases and vapors, Bureau of Mines Bulletin 627, U.S. Government Printing Office, Washington, D.C., 1965.

5. Heskestad, G., Fire Plumes, Flame Height, and Air Entrainment, in *The SFPE Handbook of Fire Protection Engineering*, 3rd ed., Sec. 2, Chap. 1, National Fire Protection Association, Quincy, MA, 2002.

6. McCaffrey, B., Flame Height, in *The SFPE Handbook of Fire Protection Engineering*, 1st ed., Sec. 2, Chap. 1, National Fire Protection Association, Quincy, MA, 1995.

7. Gupta, A.K., Fire plume: theories and their analysis, *J. Appl. Fire Sci.*, 2(4), 1993.

8. Morton, B.R. et al., Turbulent Gravitational Convection from Maintained and Instantaneous Sources, *Proc. Roy. Soc. Lond.*, 234, 1956.

9. You, H.Z. and Kung, H.C., Strong Buoyant Plumes of Growing Rack Storage Fires, 20th Symposium on Combustion, Combustion Institute, Pittsburgh, PA, 1984.

10. Kung, H.C., et al., Ceiling Flows of Growing Rack Storage Fires, 21st Symposium on Combustion, Combustion Institute, Pittsburgh, PA, 1986.

11. The author collected this data for the project reported by Sheppard, D., Fire Environment Tests Under Flat Ceilings, Society of Fire Protection Engineers, Test Report: R18476-96NK37932, Washington, D.C., July 1998. This data is available from the Society of Fire Protection Engineers.

12. Beyler, C.L., Fire plumes and ceiling jets, *Fire Safety J.*, 11(1/2), 53–75, July/September 1986.

13. Heskestad, G. and Delichatsios, M.A., The Initial Convective Flow in Fire, 17th International Symposium on Combustion, Combustion Institute, Pittsburgh, PA, 1978.

14. Tien, C.L., Lee, K.Y., and Stretton, A.J., Radiation Heat Transfer, in *The SFPE Handbook of Fire Protection Engineering*, 3rd ed., Sec. 1, Chap. 4, National Fire Protection Association, Quincy, MA, 2002.

15. Holman, J.P., *Heat Transfer*, 7th ed., McGraw-Hill, New York, 1990.

16. Mulholland, G.H., Smoke Production and Properties, in *The SFPE Handbook of Fire Protection Engineering*, 3rd ed., Sec. 2, Chap. 13, National Fire Protection Association, Quincy, MA, 2002.

17. Drysdale, D.D., *An Introduction to Fire Dynamics*, 2nd ed., John Wiley & Sons, New York, 1999.

18. Quintiere, J.G., *Principles of Fire Behavior*, Delmar Publishing, New York, 1997.

19. *The SFPE Handbook of Fire Protection Engineering*, 3rd ed., Sec. 2, Chap. 1, National Fire Protection Association, Quincy, MA, 2002.

Fire Scene Investigation: An Introduction for Chemists

2

GREGG A. HINE

Contents

0-8493-7885-0/04/$0.00+$1.50
© 2004 by CRC Press LLC

We approached the case, you remember, with an absolutely blank mind, which is always an advantage. We had formed no theories. We were simply there to observe and to draw inferences from our observations.

Sherlock Holmes
The Adventure of the Cardboard Box, 1893

2.1   Introduction

This chapter is an introduction to fire investigations. It provides the chemist with a basic overview of fire department activities, how and why investigations are conducted, fire scene safety, the collection of evidence, and a discussion of arson enforcement resources. Becoming a professional fire investigator today requires an increasing amount of training, education, and experience. This chapter does not attempt to address all aspects of fire investigation; however, it will provide the chemist with the basic tools to operate as a member of an investigative team at a fire scene and to be conversant with a local fire investigator when discussing an investigation. There are many books, publications, articles, and other resources that address fire investigations; these should be consulted for more detailed information in this field. Additionally, it should be noted that for the purposes of this chapter the discussions are relevant to investigating structure fires. Many of the same principles and procedures apply to other types of fires, such as those that involve vehicles, marine vessels, and wild lands.

2.2   Fire Investigation Issues

2.2.1   The Fire Problem

According to the U.S. Fire Administration, in the year 2000 there were 1.7 million fires reported in the U.S., and many others were unreported. The U.S. has one of the highest death rates due to fire in the industrialized world at 14.5 deaths per million population. A total of 4,045 Americans lost their lives in fires in 2000, and another 22,350 were injured as the result of fire.

The direct property loss due to fire was estimated to be $11 billion. An estimated 75,000 fires were incendiary or suspicious in nature and resulted in 505 civilian deaths and $1.3 billion in property damage. As illustrated in the U.S. Fire Administration's statistics, fire has a considerable impact on our society and takes an enormous toll in lives and property.

2.2.2   What Is Arson?

The term "arson" is commonly used today to describe a crime that involves the intentional burning of property. It originates from an Anglo-French word meaning "the act of burning." The common law definition of arson was the willful and malicious burning of a dwelling; over the years, state statues and federal laws have replaced the common law definition. Most of today's arson laws involve the intentional burning of property, not only dwellings. Statues vary from jurisdiction to jurisdiction. It is recommended that you consult local, state, or federal statutes for more details and specific language and application.

2.2.3   The Role of the Fire Department

It is the role of the municipal fire department to respond to reports of hostile fires and take appropriate action. Members of the local fire department are typically the first officials to arrive at the scene of a fire. Depending on the severity of the fire, numerous firefighting assets may arrive and participate in the operations. It is at this point that the fire investigation really begins. While not formally trained as fire investigators, firefighters can make note of the time of the fire, the fire conditions, the weather conditions, and the point of entry to suppress the fire. In addition, any suspicious or unusual activity surrounding the fire should be noted, including burn patterns, open doors or windows, alarms, unusual odors, deep-seated fire, and overall behavior and conditions.

The actions of a fire department at the scene can be separated into three distinct phases: (1) suppression, (2) overhaul, and (3) investigation. During the fire suppression phase, the first goal is to save lives; the second goal is the suppression of fire and the protection of property. In their mission, firefighters typically utilize hoses that are 1-1/2 to 2-1/2 in. in diameter to control and suppress the fire (Figure 2.1). As mentioned in the previous chapter, the application of water removes one side of the fire "triangle" — heat. After the fire has been extinguished, firefighters will search for hidden fire in walls, ceiling spaces, or other areas that are not easily accessible. This phase is termed "overhaul," and includes opening walls, pulling down ceiling materials, removing flooring, etc., to ensure that the fire has been completely extinguished. During overhaul, firefighters can unwittingly alter the fire scene

Figure 2.1 (see color insert following page 54) Firefighters advance 1-1/2 in. hose lines into a burning structure and apply water directly to the fire, removing one leg of the fire triangle and extinguishing the fire.

by removing furnishings, devices, wiring, walls, and ceiling or framing materials. Unfortunately, this alteration of the fire scene can create difficulties for the fire investigator. Depending on the jurisdiction involved, the fire scene investigation can occur in conjunction with overhaul, where the investigators are on the scene to direct the overhaul activities and to ensure the preservation of evidence.

The final phase of fire scene activities involves the investigation of the fire with the intent of determining its origin and cause. Although this is identified as the final phase, the investigation of the fire scene can actually begin during the suppression or overhaul phase. The commencement of the investigation depends largely on the time of arrival of the fire investigator or the abilities and responsibilities of the fire suppression personnel. Fire investigators will attempt to determine whether the fire was accidental or intentionally set (incendiary). Upon completion of the fire scene investigation, the property is typically released to the property owner or insurance company for further action.

2.2.4   Why Investigate Fires?

As previously discussed, fire is enormously costly to society. Fire departments across the country are required to investigate fires to determine the origin and cause. A fire occurs when a fuel comes together with oxygen and a heat source. It is the role of the fire investigator to determine how and why these

Figure 2.2   A team of fire investigators systematically examines the scene to determine the origin and cause of the fire.

factors came together and to answer the question: "Was this an accident or an intentional act?" It is not only the role of the fire department to suppress the fire but also to identify the cause of the fire.

The two primary reasons that fires are investigated are to determine what caused the fire and to identify and collect any evidence related to that cause. The purpose in determining the cause of the fire is to prevent the situation from occurring again. This is accomplished by identifying hazardous conditions or practices, product failures, or other fire causes. Once the cause is known, officials can educate the public or seek code changes in an effort to limit similar types of fires. Products suspected of causing fires can be examined more closely and modifications can be made or recalls issued in an effort to take corrective action. The second reason to investigate fires is to obtain evidence necessary to hold accountable the person or entity responsible for the cause — particularly in the case of an intentionally set fire. If the fire is determined to have been set, investigators will search for clues in the ashes in an effort to solve the crime (Figure 2.2).

2.2.5   Who Conducts Fire Scene Investigations?

This question is not always easy to answer and is sometimes unclear. While sworn law-enforcement officers typically investigate alleged or suspected crimes, this is not always the case with fire scene investigations. Persons from both the public and private sectors — often persons with varying technical backgrounds — investigate fires. Many of the public sector investigators are not law-enforcement officers but have some level of knowledge and have achieved a level of competence in fire investigation. The person conducting the fire investigation could be a volunteer firefighter with minimal training or a full-time, career investigator working for the police or fire department. In more rural parts of the U.S., a full-time fire investigator may be hours or days away from the scene. Most states have a fire marshal's office, yet these investigators typically have large geographical areas to cover and, as a result, may not be able to investigate every fire. Often it is the responsibility of the volunteer fire department to make the initial examination to determine the cause of the fire and request additional investigative resources if required.

In many cases, the property involved in the fire is insured. As a result, the private side of fire investigations can have a role in the investigation of the fire scene. Yet, this typically occurs after the public sector investigators or the fire department have concluded their investigation and the custody and control of the property is returned to the owner. Private fire investigators and insurance representatives often visit the scene and conduct an independent investigation of a fire to document the cause and collect any relevant evidence. Depending on the circumstances of the fire, insurance companies can hire other experts. These experts can include fire protection engineers, electrical engineers, metallurgists, forensic scientists/chemists, heating and air conditioning specialists, and others. Attorneys involved in investigations, whether public or private, will often visit the fire scene to make observations and discuss findings with other experts.

As you can see, the investigation of fires can be somewhat complex and not as clear-cut as other forms of investigation. Fires are investigated by a wide range of personnel, and can involve a very limited investigation or an extensive investigation conducted by highly trained and experienced members of a fire investigations unit within the fire or police department or even the local prosecutor's office. The private sector can play a large part in the investigation of fires.

2.2.6   Fire Investigator Certification Programs

Public sector fire investigators typically receive formal training in fire investigation from state and local organizations, colleges, and on-the-job training working with experienced fire investigators. The term "Certified Fire Investigator" or CFI is often used in the fire investigation community to identify an individual who has obtained a recognized level of education, training, and experience. The term, however, is sometimes loosely applied to an individual who has been certified by a state as a fire marshal or fire investigator. While the term may be variously defined depending on the area of the country, there are presently only two organizations that certify fire investigators based on a minimum set of standards for education, training, and experience: the

International Association of Arson Investigators (IAAI) and the Bureau of Alcohol, Tobacco, Firearms, and Explosives (ATF). Successful completion of these certification programs results in the designation of CFI. The IAAI certification process requires fire investigators to submit a detailed application describing their experience, training, education, and courtroom testimony. Once the minimum standards have been met and the application is approved, the candidate must take and pass a written examination administered by the IAAI. ATF also conducts a certified fire investigator program as part of its arson enforcement mission; however, the program is limited to ATF special agents only. This certification program spans a two-year period in which the CFI candidate must investigate and document 100 fire scenes with experienced fire investigators. In addition, ATF CFI candidates must attend approximately 200 h of ATF-organized training in the areas of fire investigation, fire behavior, and courtroom testimony. The candidate must complete various reading and writing assignments related to fire investigation, successfully complete two undergraduate fire science courses at a national university, and conduct a research project relating to some aspect of fire behavior or fire investigation. Upon completion of these requirements, the candidate is designated a Special Agent/Certified Fire Investigator. Both the IAAI and ATF certification programs have a recertification process that requires the investigator to maintain a level of competence by attending fire investigation training and conducting fire investigations.

2.2.7   Fire Scene Safety

The safety of the fire investigators and other investigative personnel working in and around fire scenes should be of utmost concern. Fires can cause a great deal of destruction and, as a result, dramatically impact the stability of a structure. Personnel involved in a fire investigation typically must enter the structure in an attempt to determine the origin and cause of the fire, document the scene, and collect appropriate evidence. It is paramount that these activities are conducted in a safe manner and that personnel are provided with the proper safety equipment, as illustrated in Figure 2.3. Working in and around fire scenes can be inherently dangerous, and the proper safety measures should be followed for every investigation.

Prior to entering the fire scene, the investigator should make an assessment of the exterior of the structure to evaluate potential hazards. Close attention should be paid to any nonsupported or partially supported building components (walls, floors, roofs, stairs) that may have the potential for collapse (as depicted in Figure 2.4). The location and status of all utilities (gas, water, electricity) should be determined as well as the presence of any hazardous materials (see Figure 2.5). If fire department personnel are still on the scene, they can provide an assessment of the building and point out any

Figure 2.3 When working in and around a fire scene, the proper safety equipment should always be worn to protect against exposure to contaminants, sharp objects, falling debris, and other hazards.

hazardous situations that they have found during their fire suppression efforts. In addition, the fire department may be able to monitor the environmental conditions inside the structure, including the levels of oxygen, carbon monoxide, or other contaminants, with the use of an air monitor.

Not only is the fire investigator concerned with the integrity of the structure but other health and safety issues as well. One of the greatest safety hazards at a fire scene can be airborne contaminants. With the use of synthetic materials in household and commercial products, fire investigators are frequently exposed to respiratory hazards while conducting investigations. In a typical structure fire, products containing plastics, foams, insulation, paints, and fibers are nearly always present. When these materials are involved in fire, they can liberate gases and vapors as well as aerosols, fibers, and particles. Combustion products typically found at a fire scene include carbon monoxide, hydrogen cyanide, oxides of nitrogen, and aldehydes (formaldehyde). Exposure to these contaminants can produce both acute (immediate) and chronic toxic effects.[2] Protection from respiratory hazards in the form of gas, vapor, or particulate material must be considered. Inexpensive air-purifying respirators can be worn in a fire scene with little discomfort and do not hamper the investigator's ability to examine the scene (as depicted in Figure 2.6). Since

Figure 2.4 The potential for structural collapse is an ever-present danger at fire scenes. An assessment of the condition of the structure should be conducted prior to entering.

fire investigators typically do not know all the respiratory hazards that may be present within a structure, it is important to equip the respirators with filters that protect against particulates, volatile organic compounds, acid gases, and formaldehyde. Most vendors offer these levels of protection in a single filter and these filters are readily available to the public.

Other health and safety issues at the fire scene include electrical activity, pooled water, confined spaces, biological hazards, and low lighting conditions. One of the primary safety considerations at the fire scene is the public utility service to a structure. Fire departments will typically secure the electrical service to a residential or light commercial structure by removing the electric meter. Gas service is easily controlled by the closure of a valve at the gas meter. Although the fire department may have terminated electrical activity to a structure, any electrical device or wiring should be treated as if energized until it is fully evaluated by the fire investigator. On occasion, a building may have more than one electrical service that is not apparent to the fire department or utility company and therefore not deactivated. Building occupants may have installed nonapproved wiring or may have illegally run electrical service from nearby structures; where present, these sources need to be deactivated. It is recommended that an alternating current (AC) voltage detector or similar device is utilized to detect the presence of any

Figure 2.5 The location and status of utilities (electric, gas, water) should be determined in the initial stages of the fire scene investigation. Hazards should be mitigated prior to entering.

voltage prior to handling electrical wiring. These detectors are inexpensive and provide a great deal of safety when working in and around a fire scene.

Pooled water is also a concern, especially when conducting an investigation in a basement or on a concrete slab. What looks like a shallow pool of water may actually be a deep sump pit or other opening in the floor. Water in the structure can also contain bio-hazards (sewage, infectious waste), particularly in abandoned buildings where homeless individuals may have loitered or others have utilized the structure for criminal activities.

To protect against the typical hazards encountered at a fire scene, the following minimum safety equipment should be worn or utilized:

- Hard hat or helmet
- Steel-toed boots with steel shank
- Air-purifying respirator with appropriate cartridges
- Coveralls
- Work gloves
- Eye protection
- Ear protection (if necessary)
- Flashlight

Figure 2.6 Because of the numerous respiratory hazards present at fire scenes, proper respiratory protection should be employed.

Because of the potential safety issues at a fire scene, it is recommended that two or more personnel are together at all times. No personnel on a fire scene should ever enter a structure without notifying someone. Fire department investigators typically are required to notify the department's incident commander of their presence on the scene and their anticipated activities. The incident commander often authorizes entry to the structure when all primary fire suppression activities have been concluded. Fire departments typically utilize accountability systems that document the personnel operating at a scene. These accountability systems should also include fire investigators and related personnel such as forensic scientists. The use of two-way radios by fire investigators also allows for quick notification in case of an emergency.

When departing a fire scene, all personnel should notify fire department officials that their activities have been concluded. Any soiled or contaminated clothing (coveralls, boots, gloves, etc.) should be isolated in a plastic-type bag for later cleaning or decontamination. These articles should never be taken home for cleaning as there is potential for cross-contamination. Commercial cleaning facilities should be utilized to properly clean these items.

As discussed in this section, the fire scene can be a dangerous place, and it must be treated with the utmost respect.

2.2.8   Legal Considerations

The fire scene investigator must determine the legal authority that allows entry to a property prior to conducting an investigation. It is generally recognized that fire departments have the legal authority to investigate the cause of fires for the purposes of public safety, as previously discussed. However, this right of entry is not unlimited and, according to various legal decisions, must be conducted in a reasonable period of time. The Fourth Amendment to the U.S. Constitution addresses search and seizure. Fire investigators must adhere to these standards and follow the exceptions closely.

The two Supreme Court decisions that have had the most impact on the fire investigator's right of entry involve *Michigan v. Tyler* (1978) and *Michigan v. Clifford* (1984). In the Tyler case, fire investigators left the scene and returned later that day to continue their investigation. The Supreme Court ruled that once the investigators departed, the property owner's expectation of privacy was restored. Evidence recovered by investigators in the later search was ruled inadmissible. There had been a 5-h lapse between the suppression of the fire and the initiation of the fire scene investigation. Investigators in the basement of the residence later discovered evidence that the fire was intentionally set. The Supreme Court ruled that the lapse in activity had effectively released the property back to the owner and that any further entry to the property by investigators required permission or a warrant.

While fire departments are required to investigate the cause, the investigation must occur upon suppression of the fire or in a reasonable period of time. After this reasonable time, fire investigators are required to seek other authority to reenter the property, including consent (preferably in written form) from the property owner or an administrative or criminal search warrant.

2.2.9   Scientific Method

Over the past 15 years, fire investigation has evolved into a science-based endeavor as more and more research has been conducted in the area of ignition, fire growth, and material performance. No longer can a fire investigator base his or her opinion about the cause of a fire on unsupported beliefs and mere experience. The opinion must be sound and stand the challenge of reasonable examination. Several recent court decisions have examined the methods of experts including fire investigators. These decisions have resulted in the further application and understanding of the scientific method as it applies to fire investigation. Investigators are encouraged to systematically follow the scientific method when examining the fire scene. The National Fire Protection Association's Guide to Fire and Explosion Investigations (NFPA 921) defines the scientific method as

> ...the systematic pursuit of knowledge involving the recognition and formulation of a problem, the collection of data through observation and experiment, and the formulation of testing and hypothesis.

NFPA 921 further explains how each step of the scientific method is applied to fire investigations.

2.3   Fire Scene Examination

A safe and successful fire scene examination is conducted in a reasoned and systematic manner following established procedures. The goals of any successful fire scene examination are to:

1. Determine the origin of the fire (where it began)
2. Determine the cause of the fire (the ignition source)
3. Locate, document, and preserve evidence that relates to the cause of the fire or associated criminal acts

Whether the fire scene takes minutes, hours, or days to investigate, the basic procedures are the same. Fire scene investigations typically involve three broad areas: (1) witness interviews, (2) the physical examination, and (3) forensic or engineering analysis. Since each fire is different and the circumstances surrounding the fire are also different, the degree to which each component is involved varies from fire to fire. Depending on the complexity of the fire scene, one investigator can be responsible for the entire investigation or the duties can be delegated among numerous investigators. For example, one group of investigators may be solely responsible for the scene investigation, while another group conducts all the related interviews. When this occurs, coordination between the fire scene investigators and the witness interview teams is critical so that current information and data flows between the two groups. The following sections discuss these three areas and the issues involved in fire scene investigation and the subsequent determination of the fire cause.

2.3.1 Witness Interviews

Witness interviews are conducted as part of a comprehensive fire investigation. Fire investigators seek out the information provided by witnesses or other individuals to assist them in accurately determining the cause of the fire. On many occasions, it is a witness who provides the clues to the investigators that can lead to a determination of the fire cause. A credible witness who observed the actual origin of the fire will prove invaluable to fire investigators. This is especially true when the fire has caused extensive destruction to the structure and the origin of the fire is not readily apparent to investigators. Fire investigators will attempt to locate individuals at the fire scene who either directly witnessed the fire, had knowledge about the structure, or had some background information which may shed light on the potential cause of the fire or circumstances related to the event.

In an attempt to obtain information about the fire, the following questions are typically asked of a witness:

- How did you learn about the fire?
- Who else was with you at the time you observed the fire?
- What did you do after you learned about the fire?
- What did you see?
- Where was the fire located in the structure?
- Can you describe the fire?
- Can you describe the smoke?
- Did you hear anything?
- What was the condition of doors or windows?
- Did you see any windows break during the fire?
- Did you see any person, any vehicles, or other activity around the structure either before or during the fire?
- Did you smell anything unusual?
- Did you photograph or videotape the fire?

It is always recommended that a witness be escorted back to the fire scene and interviewed at the location where the observations were made. This does not necessarily need to be done at the initial interview, but at some time during the investigation eyewitnesses should be asked to return to the scene to discuss their observations with investigators. When interviewing witnesses at the scene, investigators should walk them through their accounts of the events. This means having a witness stand in the precise location, if possible, where the fire or other event was witnessed and describe what was observed. This walk-through usually gives the investigator a clearer understanding of what a witness observed and also enables the investigator to detect any information that might be in conflict with other facts surrounding the investigation.

Many other individuals can provide a great deal of useful information to fire investigators. However, investigators must evaluate the information provided and determine if it is relevant, material, and credible. On occasion, individuals can provide false or inaccurate information for myriads of reasons. It is up to the investigator to weigh the information provided by all the witnesses against the facts in the investigation. Fire investigators typically interview all persons who have information about the scene to determine the conditions or events that may have led up to the fire. Interviews of persons in the immediate area of the scene can provide a wealth of information. These interviews, often termed a "neighborhood canvas," can include building owners, occupants, tenants, neighbors, delivery persons, postal employees, or newspaper carriers. These people can potentially provide a wealth of information relating to the fire, the structure, and its occupants.[1]

Since fires occur within a larger set of circumstances, it is useful for the fire investigator to interview all persons associated with the fire scene. This will enable the investigator to have the clearest understanding of the conditions before, during, and after the fire. This knowledge can then be compared to observations made within the scene. There are many other individuals who can provide useful information to fire investigators, including:

- Building owners and tenants
- Firefighters
- Contractors
- Insurance representatives
- Security services/alarm companies
- Local building officials
- Police officers

Some of the initial interviews that fire investigators conduct include the first-responding firefighters. The firefighters can provide valuable information relative to the location, behavior, and conditions of the fire, unusual odors, observations of unusual or suspicious activity, condition of the doors and windows, and the location and condition of victims. Contact with the first-responding firefighters is crucial in a fire investigation, particularly in cases where no eyewitness or other persons who observed the fire in its early stages can be found. In Figure 2.7, investigators interview a firefighter about the fire and suppression activities to gain a better understanding of the fire spread.

Many other individuals can provide a wealth of information regarding the fire or its cause. Contractors can provide additional information about repairs, renovations, or maintenance to the structure or its systems. Local building officials often can provide information related to building inspections and original or preexisting building permits and plans. Police officers

Figure 2.7   Firefighters can provide valuable information to investigators about the fire scene.

at the fire scene may have information related to recent activity prior to the fire or may have valuable background information on the tenants or the business.

As you can see, information gained from witnesses can be extremely useful to fire investigators. The list of sources of information cited herein provides a basic understanding of the types of details that might be useful in a fire investigation.

One issue that may be of concern during the analysis of fire scene debris relates to the presence of background contaminants or cross-contamination. This information may be useful to the forensic chemist when conducting an analysis of the debris for the presence of ignitable liquids. Could there have been gasoline or a medium petroleum distillate naturally present at the scene well before the fire? Skilled investigators must obtain the answer to this question and others like it to assist forensic chemists in their evaluation of fire scene evidence.

2.3.2   Exterior Fire Scene Examination

The ultimate goal of the fire scene examination is to identify the first fuels ignited in the fire and the source of ignition. The ignition source must be capable of causing ignition of the suspected initial fuel. For example, a single paper match, while an ignition source, is not likely to ignite a solid oak log, but that same match is easily capable of igniting a piece of newspaper. When the ignition source and the first fuel ignited are identified during the fire scene investigation, the cause of the fire may be established. Yet, the investigator must also explain how these two came together.

Generally, all fires start from a single ignition source such as an open flame (match, candle) or a hot surface, and then grow in size to room fires and to large, structure fires. Thus, it is important to focus the fire scene investigation on the early stages of the fire in an attempt to identify the initial fuel and ignition source. To accomplish this task, the fire investigator must examine the fire scene in a systematic and deliberate manner, documenting the findings along the way.

The fire scene examination typically begins from the outside of the structure and later progresses inside. The investigator should remain objective and have no preconceived notion as to origin or cause. The old saying "you can't judge a book by its cover" certainly applies to fire scene investigation. What may appear to be a likely cause in the initial stages of the investigation may not be the cause, and one must delve into the fire scene to make a true judgment about what occurred. It is for this reason that the investigation must be conducted in a systematic and objective manner and must follow the scientific method.

During the examination of the exterior, the location and description of any heat or smoke damage to the structure should be noted. The examination of the exterior and the telltale smoke and heat damage or patterns may give a general indication about the origin of the fire, as shown in Figure 2.8. Items removed from the scene by firefighters during overhaul may be found outside and require a closer examination. The condition of all doors, locks, or other points of entry should be evaluated, documented, and photographed. The utilities should be located and examined. The investigator must be aware of any items or materials that do not appear to belong in the area or seem to be out of place. For example, are any containers observed near the building? Are there ladders present, allowing access to windows and roofs? Are tools or any other articles that do not appear to belong in the area lying near a window or door? The public areas around the scene should also be examined, including pathways, alleys, lawns, parking areas, or other places that could possibly contain items related to the scene. In the case of intentionally set fires, arsonists have been known to drop containers or other materials as they depart the area. In one case, an amateur arsonist unwittingly left his wallet outside the building while rolling on the ground to stop his clothes from burning. Responding firefighters found the wallet and gave it to fire investigators who made a quick arrest in the case. As you can see, examination of the exterior is a valuable piece of the overall fire scene investigation and should be conducted in a dilgent manner to obtain all evidence related to the investigation.

Figure 2.8   Fire investigators begin the scene examination from the exterior of the structure. This examination often can provide clues about the origin of a fire.

A suitable perimeter around the fire scene should also be established at this time, if not already established by fire department personnel. The perimeter of the scene, typically established using crime-scene tape, will be used to protect the integrity of the fire scene by limiting access to only those who have official duties and whose entry is approved by the jurisdiction having authority. The size of the perimeter should be large enough to include the entire scene, any adjacent areas that are deemed relevant, and any remote locations that may include evidence or other materials relevant to the investigation. It may be necessary at times for the perimeter to be modified to include remote areas as the scene examination progresses. For example, if a gasoline container is found lying on a pathway a hundred feet from the structure, the perimeter should be expanded to include the pathway and the area up to the structure.

It is also at this time that weather conditions should be noted, as weather can play a role in fire behavior. Most important are the temperature, humidity, and wind conditions. Particular attention should be paid to the wind direction and speed as it can play a large role in fire spread, particularly in large fires.

2.3.3   Interior Fire Scene Examination

Once the entire exterior of the structure has been evaluated and a suitable perimeter established, it is time to enter the structure to begin the interior examination. At this time, the investigator should have the proper equipment to safely and successfully investigate the fire scene. The recommended minimum equipment to be used during the fire scene examination includes:

- Personal protective equipment
- Flashlight (preferably lantern style)
- Writing materials (clipboard or similar)
- Assorted small tools or multipurpose tools (screwdrivers, wire cutters, knives)
- Measuring devices (20- and 100-ft tape measures)
- Camera, film, electronic media
- Shovel or other hand tools
- Rubber gloves
- AC voltage tester

These items assist in the safe, proper, and complete examination and documentation of the fire scene. Most of the small items mentioned in the list can be easily carried in a small tool bag, waist belt, or fanny pack. Additional evidence collection equipment and supplies may be required for the investigation, depending on the nature of the fire scene.

The investigation of the interior is typically conducted in a manner that follows the fire from the area of least damage to the area of most damage. By following the damage from least to most, the investigator can attempt to trace the fire back to its origin, as typically the most fire damage will occur in the area where the fire began. This is assumed because the fire usually burns at the point of origin for the longest period of time; thus, the greatest degree of damage occurs in this area. This assumption is correct if all factors within the fire scene are roughly the same. This point is illustrated in Figure 2.9, showing the early stages of a couch fire, and Figure 2.10, which depicts the resulting damage. It is clear that the greatest damage to the couch is located nearest the lamp. As you move further away from the lamp, the damage lessens. This is the basic process that fire investigators use to trace damage back to the source or origin of the fire.

Hot gases associated with a fire flow much like a liquid, leaving a pattern or a path back to the area of origin. The size of the area of origin is relative to the scene and could be a building, room, or closet, depending on the circumstances. In the case of a fire in a large 50,000-ft^2 warehouse, the area of origin may be the northwest corner of the structure and involve 5,000 ft^2. In the case of a residential fire, the area of origin may be identified as a small bedroom on the second floor. As indicated earlier, the greatest degree of fire damage "typically" occurs in the area of origin; however, this is not always the case. The fuels and fuel arrays involved in the fire must be evaluated to

Figure 2.9   In the early stage of the fire, the halogen lamp has ignited the upholstery and foam of a couch. This is the point of origin of the fire.

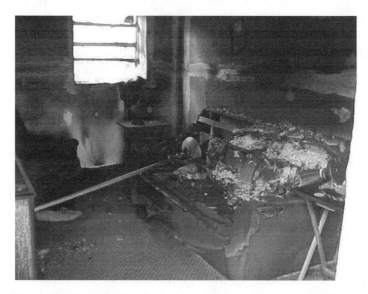

Figure 2.10   The greatest damage to the couch is apparent in the area of the lamp, which is where the fire originated. As you move further from the lamp, the degree of damage to the couch lessens.

determine their burning characteristics to ensure you are not comparing apples to oranges. As all fuels burn somewhat differently, and ventilation can affect overall fire damage, it is possible to have an area of origin with somewhat less damage than an adjoining area. For this reason it is important to identify the fuels that were present and the ventilation in that area and relate that to the observed damage.

Many factors can influence the size, intensity, and length of burn time in a particular area. Nevertheless, it is an accepted practice to initially follow the fire from the area of least damage to the area of greatest damage. During this process, however, the investigator should continually assess contents, interior finishes, ventilation, and other factors that may have influenced fire behavior and resulted in greater-than-anticipated damage outside the area of origin. As an example, firefighters arrive on the scene of a large-scale fire that has been burning for 30 min and quickly suppress the fire in the area of origin. Yet, other areas of the building may have become involved in the fire and subsequently burn for a considerable amount of time before suppressed. This scenario will likely result in the area of origin having less damage than other areas of the structure that were involved at a later time and burned longer. In this situation, witness statements and interviews with responding firefighters is critical in accurately identifying the area of origin. The fuel distribution and ventilation within a structure can also play a role in the growth of the fire. A fire that originates in an area with minimal combustible materials and then spreads to areas that have a high concentration of readily combustible materials will likely result in a higher degree of fire damage outside the area of origin.

While conducting the initial examination of the fire scene interior, all safety issues should be noted and proper precautions taken. It is at this time that photographs or video may be used to record the initial findings. If conditions change within the structure, at least the investigator has the early photographs or video to document the scene. While following the trail from the areas of least to most damage, other issues can be documented along the way. Some of these issues include the presence and location of furnishings or other contents, identification of any flammable or hazardous substances, containers, electrical wiring, doors, windows, and other openings. In the case of a fatal fire, the location of bodies, if they have not been removed from the fire scene, should also be noted. The electrical panel should be located and documented as well. The position of the circuit breakers or condition of the fuses should be noted. In the case of a potential electrical fire, circuit protection in the form of a breaker or fuse could be important to the findings.

After following the damage from least to most, and evaluating the fire behavior relative to the contents and the ventilation of the structure, the area of origin should be located. As indicated earlier, the size of the area of origin is relative to the amount of damage and the size of the structure. The area of origin can also decrease in size as additional information is obtained and the scene is examined. Fire investigators inspecting a fire involving an entire house may initially place the area of origin on the second floor. After further investigation, the investigation of the area of origin may close in on a specific second-floor bedroom, then to a portion of that bedroom.

Within the area of origin, a systematic removal of debris must be accomplished to examine the area further for fire patterns and evidence relating to the cause of the fire. This examination has often been compared to an archeological dig. As debris has accumulated as a result of the fire and destruction of the contents and structure, the area of origin is often hidden from view and any evidence of the early stages of the fire resides near the bottom of the debris. The debris within the scene must be carefully layered by hand, examined, and then removed from the fire scene, as shown in Figure 2.11 and Figure 2.12. This process is the most time consuming and requires a great deal of attention to detail on the part of the fire investigator. While systematically removing the fire debris, damage to all building materials, furnishings, electrical devices, and other contents should be documented and relevant items put aside for further examination. The investigator scrutinizes fire patterns within this area in an attempt to fully identify fire progression. The ultimate goal of this phase of the investigation is to identify the point of origin of the fire, which is described as the precise location where the ignition source and first fuel came together and burned. Sometimes the ignition source may have been destroyed in the fire or was removed from the scene by the person responsible for the fire — as in the case of a lighter removed as the arsonist flees the scene. It is up to the experienced investigator to make reasonable conclusions regarding the ignition source once all information has been gathered and the scene fully examined. The identification of the point of origin is key to the determination of the cause of the fire. In the case of an intentionally set fire, the fire investigator may discover multiple points of origin.

Once all the debris has been removed from the area and samples collected, the floor surfaces can be lightly washed with a fire hose, in a controlled manner, to remove any traces of the debris. This process can expose and highlight fire patterns and protected areas not yet observed by investigators. At this point, the remains of the furnishings can be placed back in their prefire location and examined along with the fire patterns. This reconstruction can be extremely helpful in determining the origin of the fire within a room as damage to the structure and contents are clearly visible. Documentation of this process should be made with an appropriate still or video camera.

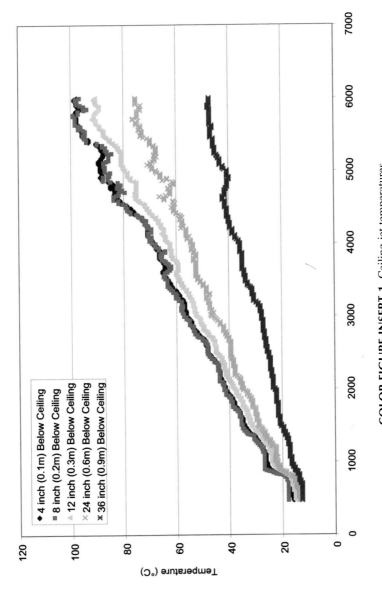

COLOR FIGURE INSERT 1 Ceiling jet temperatures.

COLOR FIGURE 1.10 Open fire in kitchen cabinets. (Courtesy of Special Agent/CFI Michael R. Marquardt, Federal Bureau of Alcohol, Tobacco, Firearms, and Explosives.)

COLOR FIGURE 1.11 Open fire in a living room, showing the fire plume and hot smoke layer. (Courtesy of Special Agent/CFI Michael R. Marquardt, Federal Bureau of Alcohol, Tobacco, Firearms, and Explosives.)

COLOR FIGURE 1.12 Fully developed fire in a bathroom. The picture was taken from outside the building through an open window. (Courtesy of Special Agent/CFI Michael R. Marquardt, Federal Bureau of Alcohol, Tobacco, Firearms, and Explosives.)

COLOR FIGURE 2.1 Firefighters advance hose lines into a burning structure and apply water directly to the fire, removing one leg of the fire triangle and extinguishing the blaze.

COLOR FIGURE 2.13 Fire patterns are the visible or measurable physical effects that remain after a fire, as seen on this exterior door. A fire was set directly in front of the door, causing the visible damage.

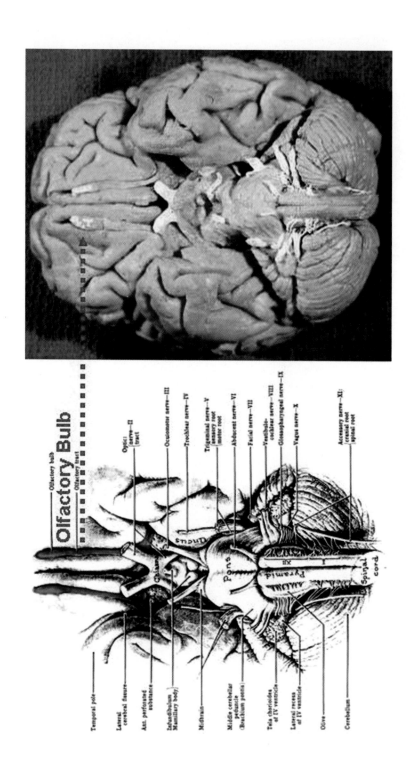

Olfactory Bulb

Olfactory bulb
Olfactory tract

Optic nerve—II
tract

Oculomotor nerve—III

Trochlear nerve—IV

Trigeminal nerve—V
sensory root
motor root

Abducent nerve—VI

Facial nerve—VII

Vestibulo-
cochlear nerve—VIII

Glossopharyngeal nerve—IX

Vagus nerve—X

Accessory nerve—XI:
cranial root
spinal root

Pons
Pyramid
Spinal cord
Midbrain
Olive

Temporal pole

Lateral
cerebral fissure

Ant. perforated
substance

Infundibulum
Mamillary body

Midbrain

Middle cerebellar
peduncle
(Brachium pontis)

Tela chorioidea
of IV ventricle

Lateral recess
of IV ventricle

Olive

Cerebellum

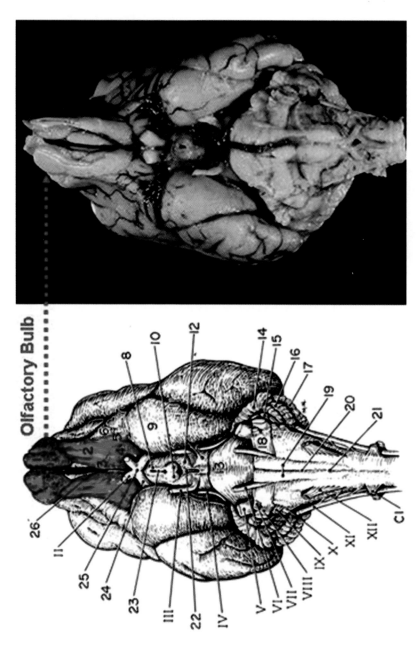

COLOR FIGURE 3.2 A comparison of human and canine brains demonstrates the significantly larger olfactory track and bulb for the canine. (Sources: *Grant's Atlas of Anatomy*, Williams & Wilkins Co., Baltimore, 1978; http://www.sci.uidaho.edu/med532/cranialnervestartpage.htm; *Miller's Anatomy of the Dog*, 3rd ed., W.B. Saunders, Philadelphia, 1993; http://cal.vet.upenn.edu/neuro/server/lab4frameset.html.)

COLOR FIGURE 3.1 Accelerant Detection Canine "Villain" indicates presence of lamp oil in five separate areas of a room with all validated by the lab, corroborating the dying statement of the victim. (Used with permission from Bill Whitstine, Florida Canine Academy, 19 Marshall St. Safety Harbor, FL.)

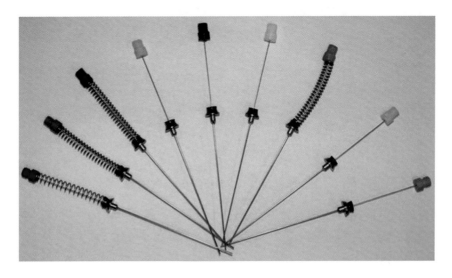

COLOR FIGURE 3.5 Examples of different SPME fibers available.

Figure 2.11 Fire investigators must search the scene in a systematic manner to locate items that can assist in the determination of the origin and cause of the fire.

Figure 2.12 Fire investigators must carefully examine items found at the fire scene that may have a bearing on the cause of the fire. If further analysis is appropriate, the item can be collected and preserved at the fire scene.

2.3.4   Fire Patterns or Fire Indicators

The principal objectives in the determination of the origin and cause of a fire is the recognition, identification, and analysis of fire patterns. NFPA 921 defines fire patterns as the visible or measurable physical effects that remain after a fire (see Figure 2.13). These effects represent the history of the fire, as it is recognized that fires cause predictable patterns on materials as they burn.[2] Since fires burn at or near the point of origin longer than at other places, all things being equal, then the most destruction should be at that point. Fire investigators use these patterns as pointers to trace the path of the fire back to its origin since gases from combustion flow like a liquid and will follow the path of least resistance around obstructions in an upward manner. Further examination of the scene can be focused in the suspected area of origin once the fire patterns or indicators have been identified. However, fire patterns can be cumulative and thus result in multiple patterns being overlaid, one atop another, as the fire progresses, other fuels become involved, and ventilation conditions change. It is the responsibility of the fire investigator to examine these patterns and assign them value as appropriate. With the proper examination of fire patterns, the investigator can trace the fire back to its origin. To do this, the investigator must fully understand the physics and chemistry of fire and the modes of heat transfer: convection, conduction, and radiation.[3]

Often, there are many patterns or indicators that can be identified at a fire scene, some of which may be of value in the investigation. These patterns or indicators are not absolute and can be created in different ways. For example, the finding of thermal damage or a burn pattern on a combustible floor as shown in Figure 2.14 can be the result of ventilation, radiant energy from a nearby flame, hot gases, dropping or falling materials that burn on the floor, or the burning of an ignitable liquid or other flammable substance. The investigator may never know which event or series of events caused the fire pattern on the floor; yet, the observed damage cannot be dismissed, and must be noted and compared to other patterns or indicators that are observed at the fire scene. It is the culmination of fire patterns or indicators at the fire scene that the investigator uses to identify the origin or cause of the fire. No single indicator can be used to the exclusion of the others.

Patterns that are typically observed at fire scenes include "V" patterns, lines of demarcation, low burns and penetrations, charring (often called "alligatoring"), clean burns, and trailers. These patterns can be readily apparent to the casual observer, or hidden from view and apparent only upon removal of fire debris. Therefore, a comprehensive fire scene examination involves the systematic removal of debris so the investigator can fully evaluate the scene, the fire patterns, and the damage. It should be noted, however, that the formation of fire patterns is the subject of ongoing research to

Figure 2.13 (see color insert following page 54) Fire patterns are the visible or measurable physical effects that remain after a fire, as seen on this exterior door. A fire was set directly in front of this door, causing the visible damage.

quantify the factors involved in pattern development. Following are brief descriptions of the some of the more-common fire patterns.

2.3.4.1 V and Hourglass Patterns

As fires burn upward and outward from a fuel source due to buoyancy, they usually leave distinct patterns in the shape of a V, typically referred to as a "V" pattern. These patterns are usually apparent on vertical surfaces such as walls that are directly adjacent to a burning object. The apex of the pattern will be located at the fuel source. The pattern then widens as it spreads up and out, away from the fuel source. In Figure 2.15, a V pattern can be seen across the front of an apartment building. The lowest point of the pattern is on the second floor landing where the fire originated. V patterns can be large, as shown in Figure 2.15, or much smaller and visible on interior surfaces. This pattern is one indicator to be used in the determination of the origin of the fire. Recent analysis has been conducted to rebuff some myths regarding the angle of V patterns. Studies have shown that the width of the angle in a V pattern is associated with the rate of heat release of a material and the

Figure 2.14 Penetrations to floors during fires can be the result of radiation, ventilation, falling materials or the burning of an ignitable liquid on the floor surface. Samples of the wood flooring may be collected at the scene for further laboratory analysis.

Figure 2.15 A "V" pattern is evident on the front of this structure. The fire originated on the second floor and spread up and out, involving the third floor and attic.

length of time the material is burned.[4] Patterns can also be in the shape of an hourglass, particularly when a pool of liquid fuel burns adjacent to a vertical surface, as shown in Figure 2.16.

Figure 2.16 An hourglass pattern is formed from a burning ignitable liquid on the floor adjacent to the wall.

2.3.4.2 Lines of Demarcation

Lines of demarcation are the visible patterns or borders that delineate regions affected by heat and smoke from adjacent unaffected or less affected regions.

Lines of demarcation can be in many forms and are used by fire investigators to assess the smoke or fire progression within a structure. The lines or patterns are created by a thermal insult to an object or during the deposition of combustion products. In Figure 2.17, the pattern left by the smoke within the room is clearly evident and a line of demarcation is seen on the walls. This pattern is helpful in assessing which windows or doors were open within a room and in evaluating witness statements regarding the smoke layer. Lines of demarcation or surface effects can also be seen on any metals in the form of oxidation, discoloration, or melting. As a metal is exposed to increasing temperatures, it begins to exhibit patterns relative to the thermal exposure. These patterns can be used to evaluate the direction or location of the fire and the intensity of the exposure. In Figure 2.18, the lines of demarcation and effect on the metal surface of the device are clear. These effects

Figure 2.17   The line of demarcation separating the smoke layer above from the fresh air below can clearly be seen on the wall surfaces.

Figure 2.18   The line of demarcation on this metal cabinet was caused by extensive thermal heating within the unit. Clean paint can be seen on the lower (cooler) portion of the unit.

were caused by the overheating of the device, which subsequently ignited nearby combustibles, causing a fire. Investigators used these patterns to confirm the cause of the fire, as it was determined that the patterns were the result of internal heating of the unit and could not have been caused by external heat exposure.

2.3.4.3   Low Burns and Penetrations

The lowest point of burning observed at a fire scene should be examined closely as a potential point of origin. Any penetration in the floor (as shown in Figure 2.14) should be evaluated to determine its cause. While penetrations in the floor and associated low burning can be the result of the burning of an ignitable liquid, the patterns can also be caused by structural collapse, radiation, or the pooling or falling (drop-down) of burning materials. These areas are often the locations where flooring, carpet, and fire debris samples are recovered by investigators for forensic analysis.

2.3.4.4   Charring

Charring to wood materials is commonly found at fire scenes. The amount and depth of charring is commonly used by investigators to evaluate fire spread, intensity, and duration of the fire. While the rate of wood charring has been quantified in laboratory experiments, the use of a defined rate of charring for fire scenes is not appropriate. Since the rate of charring is dependent on the intensity of the fire, duration of exposure, species of wood, and moisture content, an evaluation of charred wood for the purposes of determining an accurate time of exposure may not be reliable for a fire scene investigation. However, the comparison of charring depths in various locations in a fire scene may be reliable in determining relative time of exposure, assuming the wood species are the same. In Figure 2.19, the charring to the floor joists is quite apparent. Two of the joints are extensively charred to the point of collapse, and as you move away, the charring lessens. This damage assessment assists the investigator in determining which joint was involved in the fire for the longest period of time. This information is helpful in determining the point of origin for the fire.

2.3.4.5   Clean Burn

A clean burn to the surface occurs at a fire scene when a surface is exposed to direct flame impingement. The direct flame contact causes the soot deposits to be burned away, leaving a clean area. The clean burn can vary in size, depending on the size of the localized flame. These patterns can assist fire investigators in identifying the location of burning materials and can sometimes lead to the origin of the fire. In Figure 2.20, a clean burn area can be

Figure 2.19   The examination of wood structural members can aid in locating the origin of the fire. The fire has extensively damaged the joists in the middle of the photograph.

Figure 2.20   A clean burn occurs when flames or intense heating burns off the soot in a localized area. This pattern was caused by a container of ignitable liquid burning near the wall surface.

Figure 2.21 A trailer of gasoline is used to spread the fire across the floor of the room from the doorway. This is a common technique used by arsonists.

seen on a wall. An ignitable liquid and a small container were burning on the floor in the area of this pattern.

2.3.4.6 Trailers and Pour Patterns

"Trailer" is a term used by fire investigators to describe a combustible material or ignitable fluid intentionally placed to spread fire from one location to another. The pattern resulting from an ignitable liquid trailer is often called a pour pattern. The telltale signs of a trailer can sometimes be observed at a fire scene; however, this is largely dependent on the overall degree of damage and the trailer materials used by the arsonist. In a postflashover environment, the persistence of ignitable liquid patterns is less identifiable.[4] In Figure 2.21, a trailer of gasoline was used to spread the fire across the floor of this room. The gasoline was first poured on a desk and then trailed out of the room

Figure 2.22   A pour pattern caused by a burning ignitable liquid can clearly be seen on the floor. As a fire grows within a structure, these patterns may become more difficult to locate.

to the doorway, where it was ignited with an open flame. During the scene examination, the location of the trailer may not be apparent in the fire debris. However, with the removal of the debris and cleaning and washing of the floor, the pattern may become apparent. In Figure 2.22, a pour pattern, caused by the ignition of an ignitable liquid, is clearly visible on the flooring. In this case the liquid burned, but there was no fire spread beyond the trailer.

If the use of an ignitable liquid trailer is suspected at a fire scene, an accelerant detection canine team should be utilized, if available, to pinpoint the location of an ignitable liquid. Ideally, the use of the canine team would occur in the early stages of the scene investigation and again upon completion of the investigation.

2.3.4.7   Scene Documentation

As with any investigation, the fire scene must be documented to record the findings of the fire investigator. Documentation can involve note taking, report writing, photography, sketching, or diagramming. The tasks provide a means of accurately and contemporaneously recording the findings of the investigator so that they can be recalled at a later time for administrative, civil, or criminal proceedings.

Photography is the easiest way to accurately document a fire scene. The use of quality photography equipment with a separate flash unit is recommended. This equipment can be an SLR 35-mm camera, a digital camera, or a video camera.

Because fire scenes by nature are typically dark with black surfaces, photography can be somewhat challenging. This is why a separate flash unit is important when photographing a fire scene. Portable lights from fire department apparatus or the fire investigations unit can be useful in lighting a scene as well. Consideration should also be given to the use of a wide-angle lens.

When taking photographs at a fire scene, it is important to record the subject of each individual photograph. Upon later review of fire scene photographs, many may look the same. Therefore, it is important to document each image in a log identifying the date, time, and subject matter of the image. Also, the first image on each roll of film or digital media should be a title sheet identifying the agency, date, location, photographer, and case number if known. This information will help in later cataloging numerous rolls of film, compact disks, and other storage media with numerous fire scene images.

A video camera may also be useful in the investigation. This type of recording is valuable when briefing others who did not visit the fire scene as it provides a better overview and typically can more easily orient the observer than a group of photographs. When documenting with a camera (film, digital, or video), it is important to methodically document the scene rather than jump from one area to another.

Usually, the scene should be documented from general to specific. In other words, the scene should be photographed first from a distance and should include landmarks, street signs, or other reference points. Once this has been accomplished, the photographer can move into the medium-range photographs and then progress to more specific areas. It is also useful to photograph the fire scene from above. This may record fire damage or evidence that was not visible from ground level, giving the investigator a different perspective on the scene. Viewing the scene from above is particularly useful in large fires and can aid the investigator in identifying areas of greater damage. Depending on the size of the scene, the overhead view can be accomplished with the use of a ground ladder, a ladder truck, or an aerial platform, as shown in Figure 2.23. To get a full view of extremely large scenes, photographs from a helicopter may be more appropriate.

It is also recommended that a diagram or sketch be drawn to reflect the geometry of the involved structure. Accurate measurements that identify the overall size of the structure, including ceiling heights and window and door

Figure 2.23   An aerial view of the fire scene can provide a beneficial perspective for the investigator. This can be accomplished with the use of a fire department apparatus, a nearby building, or aircraft.

openings, can be valuable information. This drawing can be very useful as a demonstrative tool when briefing others on the fire scene. Sketches can be hand drawn, as long as they are neat and clear. However, many fire investigators today utilize computer-based architectural drawing programs, which can produce professional-quality diagrams. Many of these drawing programs are commercially available and are fairly easy to use.

Note taking is an important method of contemporaneously recording observations and other information, and can be accomplished with pen and paper or with a small cassette recorder. Either method will work to accurately record activities at the fire scene. At some later time, the notes can be assembled with other documentation and a detailed report can be authored. For later identification, each page of notes should bear the author's name and signature, along with the date the notes were recorded.

2.3.5   Forensic Analysis

Forensic analysis to assist fire investigators is becoming more and more common. This analysis typically involves the traditional laboratory analysis of fire debris, but can also involve other forensic disciplines such as tool-mark, fingerprint, trace evidence, DNA, pathology, and engineering. A dialogue or link

Figure 2.24   Fire investigators often work side by side with engineers and other technical experts to determine the cause of a fire.

between the fire investigator and the forensic chemist or engineer is extremely important to the successful evaluation and analysis of evidence. The fire investigation is always better served as the fire investigator, forensic chemist, and forensic engineer understand and learn more about each other's roles, methods, and techniques. This increased level of awareness can be accomplished through an ongoing dialogue between the parties regarding a specific investigation as well as joint training sessions and meetings. Fire investigators have often benefited from presentations provided by forensic chemists during annual training sessions that relate to the collection and preservation of evidence and subsequent laboratory examination. Forensic chemists and engineers often benefit from participating in actual fire scene investigations and working side by side with fire investigators. Through this partnership all the parties are able to provide a better level of service, as well as to develop a better understanding of the other's duties and responsibilities.

The assistance of fire protection engineers, mechanical and electrical engineers, and fire scientists has increased in the fire investigation field. Engineers who are well versed in combustion, fire behavior, material performance, electrical systems, and fire codes are providing an ever-higher level of expertise to fire investigators (Figure 2.24). In addition, universities, research facilities, and federal agencies such as the National Institute of Standards and Technology, the ATF Fire Research Laboratory, and the Consumer Product Safety Commission are providing technical support to fire investigators in areas such as fire dynamics, product failures, and forensic fire reconstruction.

Fire dynamic calculations and mathematical fire modeling have become more advanced in recent years. The most recent fire model, developed by the National Institute of Standards and Technology, is a computational fluid dynamics model known as the Fire Dynamics Simulator (FDS). This model is linked to an animated, three-dimensional computer program (Smokeview) that aids the user in visualizing fire development and spread within a structure. State-of-the-art computer models like FDS can be used to predict fire phenomena such as time-to-flashover, gas temperatures, smoke concentrations, and sprinkler activation time. Over the past few years, fire investigators have relied on engineers to use models such as FDS to assist them in understanding fire behavior in a particular investigation or evaluate statements made by a witness or defendant regarding fire behavior. However, these models require a great deal of expertise to use effectively and to understand the science behind the model. As with any computer program there are limitations and they are not suitable for all investigations.

2.3.6   Fire Cause Classification

Once the fire investigator has reviewed all relevant facts and information surrounding a fire, the cause of the fire will be classified. Except in the most clearly defined circumstances, the cause of the fire should be based on the presence rather than the absence of evidence. The cause of a fire is generally classified as accidental, natural, incendiary, or undetermined.[4] If the cause cannot be determined, the fire should be classified as undetermined. The cause can be undetermined for many reasons and may be due to the degree of damage to the structure, lack of witness information, or other physical evidence. The classification of undetermined may change at some later time if additional relevant information is developed. The determination of any fire cause, however, must be based on credible information and facts. While some investigators have used the classification of "suspicious", this classification is discouraged because it is not an actual description of the fire cause. The following is a brief definition of the most commonly recognized fire classifications.

2.3.6.1   Accidental

This classification encompasses situations that generally do not involve direct human involvement, such as fires caused by appliance failure, electrical wiring, or other nonhuman causes. However, an accidental classification can be used in instances that encompass noncriminal human involvement. For example, a homeowner burning leaves may inadvertently cause a secondary fire in some nearby brush. While the second fire may be caused by negligence, it is still accidental in nature.

2.3.6.2 *Natural*

This classification encompasses fires that are typically identified as acts of God, such as fires related to lightning strikes, earthquakes, etc. No human involvement is linked to the natural fire classification.

2.3.6.3 *Incendiary*

These fires are situations that are intentional, malicious, and are started by direct human intervention. They are criminal in nature and are often classified by law enforcement authorities as arson.

2.4 Collection and Preservation of Evidence

Evidence can be anything that furnishes proof and assists in supporting a theory. In the case of a fire scene investigation, evidence is typically used to support the cause of the fire or other issues related to the fire scene. Fire investigators should attempt to protect and preserve the fire scene and its contents as much as possible in an effort to properly identify the prefire conditions. This is why the establishment of a controlled-access perimeter around the scene is important. The entire scene should be protected as evidence until the completion of the fire scene examination as the determination of the cause of the fire is generally not known until the end of the investigation. Items of evidence are often found at a fire scene and include fire patterns and artifacts from the initial fuel or ignitions source.[3]

Should fire investigators suspect that ignitable liquid was used to promote the rapid growth and fire spread within a building, samples of materials or debris should be collected for laboratory analysis to detect the presence of any unconsumed ignitable liquids (accelerants). Examples of other items which might be collected or documented at the fire scene include portions of a door and lock that indicate forced entry, containers of suspected ignitable liquids, tire or foot impressions, tools, documents, and blood.

As each fire scene is unique, it is the responsibility of the trained fire investigator to determine what constitutes evidence, and then make the proper arrangements for the collection and preservation of these items.

The determination of what constitutes evidence and the need for the collection of the items can change depending on the responsibility and role of the investigator. In the case of an accidental fire caused by a product failure, the government or public sector investigator may choose not to collect the suspect product from the fire scene but, rather, defer to the insurance investigator for collection of the item, subsequent analysis, and potential civil litigation. The government fire investigator, however, should always collect evidence related to criminal activity.

The evidence most frequently collected from the scene of a suspected incendiary fire is debris and other materials such as flooring, carpet, baseboard, and pieces of furnishings. These items are collected for later examination for the presence of an ignitable liquid. Information developed by fire investigators from witnesses and the fire scene examination will generally lead to a determination as to the origin of the fire. As mentioned earlier in this chapter, the fire investigator must then seek to identify the specific cause of the fire within the area of origin. If the cause of the fire is suspected to be incendiary in nature, the fire investigator may possibly collect samples from the scene to determine if an accelerant was used. However, arsonists often utilize available materials such as paper, cardboard, and other lightweight items to initiate and accelerate the spread of the fire. Thus, the use of an ignitable liquid may not occur in all incendiary fires. It is up to the trained fire investigator to be aware of any telltale signs relative to the use of an ignitable liquid. Some of these signs could include abnormal fire spread, intense localized fire damage, the presence of flammable liquid containers, the odor of petroleum products, a visible sheen on the surface of pooled water and burn patterns consistent with the use of an ignitable liquid.[3]

2.4.1   Cross-Contamination Issues

The potential for contamination of evidence during the collection of samples from the fire scene must be considered by the fire investigator well before the actual collection of samples is conducted and even before arriving at the fire scene. The evidence samples collected from the fire scene should not be contaminated with any substances either prior to or during the collection process. It is for this reason that the use of gas-powered equipment within a fire scene should be limited and all evidence collection equipment and containers be maintained in an appropriate manner. Unfortunately, it is not always possible to control the activities of the fire department personnel during fire suppression activities. Therefore, a fire investigator should make an accurate accounting of the use of power equipment in the fire scene and evaluate the potential for cross-contamination of samples collected from the fire scene. Any concerns relating to this issue should be fully discussed between the fire investigator and the forensic chemist.

Contamination can occur through the use of tools, evidence collection equipment, and evidence containers, clothing, and footwear. Therefore, it is recommended that all items used in connection with the collection of evidence at a fire scene be thoroughly decontaminated prior to use. Fire investigators must also consider potential background contamination that may be naturally present at the fire scene. Medium petroleum distillates are often

Figure 2.25 Metal cans are commonly used to collect and preserve fire debris for laboratory analysis. A lined paint can will limit the degradation of the container. All samples collected from the scene should be properly documented.

used as a carrier for insecticides, flooring adhesives contain solvents and various commercial cleaning supplies are petroleum based. The collection of a comparison sample may be useful in some situations so the forensic chemist can determine if the sample in question has some potential background contamination. Comparison samples are defined as materials that are not suspected to contain any contamination and accurately represent the pre-fire condition of the material to be tested. The comparison sample is typically collected as close to the original sample as practical, but ideally in an unburned area and not exposed to water. If this is not possible, then a sample should be taken in an area where the presence of an ignitable liquid is not suspected.

Comparison samples are not required for routine identification of common ignitable liquids.[6] The fire investigator must determine the potential for cross-contamination through witness interviews and observations made at the fire scene. The forensic chemist can provide further guidance relative to the use of comparison samples.

2.4.2 Collection Procedures

Samples of fire debris or the material suspected of containing ignitable liquid residue are generally placed in clean, unused metal cans with a friction-fit lid (often called a "paint can") as shown in Figure 2.25. These metal containers are the same type used for the retail sale of paints. They provide a secure and convenient way to collect and preserve fire scene evidence that may possibly contain volatile residues. Because by nature fire debris tends to be wet, the use of Teflon™-lined cans is recommended over unlined cans for the collection of fire debris at fire scene. The lined cans will limit the potential for rust, keeping the sample intact. The sample taken today at a fire scene

Figure 2.26 Small, inexpensive garden tools such as hand trowels and rakes work well when attempting to collect fire debris for analysis. As with any tool used within the fire scene, it should be properly decontaminated prior to use in collecting samples.

may not be introduced in court for many months or years. The ability to present a pristine evidence can provides a much-better image to the court than a rusted and degraded can that is leaking its contents.

The use of an accelerant detection canine (ADC) should be considered when evaluating the cleanliness of evidence containers and collection equipment. The ADC team can inspect all evidence collection material before it enters the scene to confirm that no cross-contamination exists. If possible, it is recommended that an ADC team be used by fire investigators before they ever respond to a fire scene to assist in inspecting the evidence containers. Once the inspection is complete and the evidence cans are determined to be clean, they should be closed, sealed with evidence tape, and marked with the date that the can was inspected by the ADC team. This process allows for the fire investigator to bring a preinspected evidence container into the fire scene.

Disposable tools are often useful in the collection of fire debris evidence. Hand shovels, rakes and trowels used in gardening are inexpensive and work well when collecting small amounts of debris for analysis, as depicted in Figure 2.26. The shovel is rigid enough to scrape and pry debris that may have adhered to the substrate. These tools can be disposed of at the conclusion of the fire scene investigation or decontaminated.

When collecting debris samples from the fire scene for subsequent laboratory analysis for the presence of ignitable liquids, the following procedures should be followed:

- Wear new, clean and unused disposable gloves during the collection of each piece of evidence.
- If possible, only one person should handle and package the evidence to eliminate potential chain of custody issues.
- Use a clean (decontaminated), unused tool in the collection of evidence or utilize the can lid to assist in the recovery process.
- Use an evidence marker to identify each item of evidence collected; photograph the evidence collection area with and without the evidence marker in place.
- Photograph each step of the collection process. Make sure to include one photograph that shows the relative location of the sample collected within the structure.
- After the sample is collected, seal the friction lid of the can, making sure that no debris rests in the groove of the can. Use a rubber mallet to seal the can. The use of a hammer is not recommended, as it can damage the can and disrupt the seal.
- Write the date, time, location, item number, description, and collector's name on the metal can using a permanent marker; additional information can also be added as appropriate.
- After the evidence container is sealed, new disposable gloves should be worn by the person collecting the next item of evidence should don new disposable gloves.
- Dispose of the contaminated tools or properly decontaminate at the conclusion of the fire scene examination.

As indicated, the collection of evidence at the fire scene must be conducted in a methodical manner while fully documenting the process. The exact location of the collection site within the structure can be very important, especially when presenting the information in court. Photographing the entire process can pay dividends in the end. It is also recommended that all items collected be recorded on an evidence log that corresponds to the information recorded on each evidence can. Care should also be taken on the log to record the precise location where the sample was collected within the fire scene. The location of the evidence collection sites can also be recorded on a fire scene diagram or sketch.

2.5 Summary

The goal of any successful fire scene examination is to accurately identify the first fuel ignited and the ignition source that caused the fire. Many factors enter into this process, and this is often not a simple task to accomplish. As represented in this chapter, fire investigations can be complex endeavors that involve many different disciplines. Witnesses must be interviewed, the scene examined, and analysis conducted. No one factor or indicator can be used to determine the origin or cause of the fire. It is the analysis of all relevant data that leads the fire investigator to an accurate determination of the fire cause. Once all the information is assembled, the fire investigator must make a reasonable judgment as to the cause of the fire and answer the question: "Was this an accident or a crime?" As with many specialties, it can take years for a person to obtain the requisite experience, education, and training to be a successful fire investigator. This chapter provides the forensic chemist with a general overview of fire scene investigations so he or she can participate as a member of a fire investigative unit or adequately discuss a fire scene examination with a fire investigator. Many other texts and publications exist that can provide additional information on fire investigations and related activities and should be consulted for additional information on this subject.

References

1. Huff, T., The Neighborhood Investigation, National Center for the Analysis of Violent Crime, Quantico, VA, 1998.

2. DeHann, J., *Kirk's Fire Investigation*, 5th ed., Prentice Hall, Upper Saddle River, NJ, 2002.

3. NFPA 921: Guide to Fire and Explosion Investigation, National Fire Protection Association, Quincy, MA, 2001.

4. Federal Emergency Management Agency, United States Fire Administration, USFA Fire Burn Pattern Tests, FA 178, 1997.

5. Department of the Treasury, ATF, Special Agents Guide to the Forensic Science Laboratory, September 1999, ATF P 7110.1 (09-99).

Detection of Ignitable Liquid Residues in Fire Scenes: Accelerant Detection Canine (ADC) Teams and Other Field Tests

3

KENNETH G. FURTON
ROSS J. HARPER

Contents

3.1 Introduction

Once at the fire scene, it is the role of the forensic investigator to determine the seat, or origin, of the fire. This part of the investigation is essential to establish the cause of the fire as well as how it developed and spread throughout the entire fire scene. In turn, examination of the origin of the fire will facilitate the progressive determination of the cause of the fire, whether accidental or malicious. To best determine the likelihood of a willful fire-raising event, or act of arson, it falls to the forensic investigator to collect samples for evidence to assist the investigation, much like any other crime. These samples must then be analyzed at a laboratory for the presence of any accelerants or ignitable liquid residue (ILR). Initial investigations involve visual cues about the fire source, followed by either instrumental or biological tools to help further pinpoint where the fire may have begun and if there are any residues remaining that could indicate arson. Accelerant Detection Canine (ADC) Teams are increasingly employed at fire scenes to provide assistance in the search for the presence of ILR. There are, however, various instrumental techniques, in addition to the basic interpretation, through experience, of the fire scene itself, which can be applied to achieve the same result. Furthermore, the importance of correctly identifying the location of ILR at a suspected arson scene is twofold. The presence of ILR at the scene can only be confirmed by subsequent instrumental analysis at a laboratory. By correctly isolating the area suspected to contain ILR, the number of samples that must be collected can be severely reduced, lessening the amount of work required by the laboratory.

Having detected ILR at the scene, it is also important to be able to accurately place that residue within the crime scene. If the ILR is detected at the seat of the fire, it can be proposed that an accelerant may have been applied to set the fire. If the ILR is detected away from the seat of the fire, then it cannot be proved that an accelerant was used to willfully start the fire, since it would be argued that the ignitable liquid was present elsewhere in the fire scene innocuously and not present at the origin. A major concern, which the investigator must address, is that of personal safety. It is common for serious structural damage to occur during the process of a fire, which may render areas of the building unsafe to enter. The fire scene investigator should always be equipped with protective gear, including safety helmet and shoes. In older buildings, it quite possible for contamination of the scene with asbestos or other toxic material, and in addition to pyrolysis fumes, this may necessitate the use of respiratory equipment. Before entering the building, the investigator should seek the advice of the relevant authorities, in particular the fire fighters themselves. The investigator should also be prepared to confine the investigation to areas declared safe, and in doing so avoid part of the crime scene. When using a canine search approach, consideration of safety must be extended to include that of the dog.

3.2 Visual Inspection of Fire Scene

Visual methods of fire scene analysis rely on the experience of the fire scene investigator to identify and pick out important areas within the debris to give certain clues. The investigator must consider every aspect of the scene: the floor, walls, ceiling and room contents individually and as a whole to draw together appropriate conclusions. There are many different visual aids and investigative avenues to follow when inspecting a fire scene.[1,2]

Methods that rely on the hypothesis that the fire is most likely to have burned the longest at the origin are referred to as time/temperature-dependent techniques. This hypothesis is not always valid, but assumes that the origin will have burned the longest and achieved the highest burning temperature.[3]

Techniques to determine the length of time of burning from the depth of char can be applied to different materials observed at the scene. There is a widespread acceptance that wood chars at a steady rate; however, the depth of charring of wood is shown to be a function of the flux of radiant heat and the length of time that the wood was exposed. It follows that the hotter the fire environment and the greater the length of time of exposure, the wood will undergo more excessive charring. Recall that there is a sharp thermal gradient between the ceiling and floor areas within a fire scene, and thus care must be applied with this technique, as wood at a higher level will char preferentially over a lower level surface. Comparisons of similar wood types are essential. Clearly a soft wood such as pine or maple will char significantly faster than a hard wood such as mahogany.

Correspondingly, damage to glass can be interpreted for time and heat exposure. Glass melts at over 850°C, but distortion is observed at temperatures upwards of 700°C. A light bulb is simply a glass ball filled with gas, and as the heat flux from a fire radiates toward the light bulb, the side of the bulb closest to the fire will heat and thus soften preferentially. As the gas inside the bulb heats, it will expand, distorting the shape of the light bulb. Therefore, distortions in light bulbs will usually point in the direction of the heat source of the fire scene, a useful indicator to locate the main fire area. Elsewhere, examination of the edges of broken glass fragments can indicate whether the glass was broken before or during the fire, characterized by softened edges, or after the fire by the sharper glass shards. Likewise, rolled steel joists, when subjected to heat and stress, can deform to leave visible evidence. The degree of distortion is a function of the radiant heat flux experienced and the applied stress or the load that the steel beam was bearing. Significant distortion or oxidation is a clear indication of high temperature damage. Many materials change their appearance upon exposure to heat or flame, whether it is wood charring, metal oxidizing, or paint cracking. The direction of the heat source can often be determined by careful examination of the degree of damage across the subject surface.

Geometrical techniques consider the spread of the fire, applying common rules and physics to the fire scene. The pyrolysis gases naturally rise due to their buoyancy, and the thermal gradient of a fire scene tends the fire to spread in an upward direction. It must be considered, however, that destruction of the scene and falling, burning debris will spread the fire downwards, giving many directional indications. Given the natural tendency for fire to spread upward, the lowest point of burning is often assumed to be indicative of the fire seat. This is clearly not always the case as downward fire development has also been observed, and falling debris may start other fires that can be mistaken for the origin. Difficulty may also exist when the fire burns through a floor to the level below, again confusing the lowest point.

Smoke records, the study of soot deposition on surfaces within the fire scene, is considered a developmental method, as this evidence develops as the fire spreads. Smoke particles will bind preferentially to cool rather than hot surfaces, and rough surfaces offer more traction than smooth polished ones. Soot is flammable such that smoke deposition will be observed around the cooler edges of a fire scene, but not at the origin, where temperatures would be sufficiently high to volatilize the soot and combust it.

In addition to giving indications as to the position of the seat of the fire, smoke deposits can assist in determining the position of objects before and during the fire. Clearly, this method of analysis is dependent on time and temperature effects.

One must also consider the effects of smoke. Smoke deposits favorably on cool, rough surfaces rather than hotter or smoother surfaces. The chemical composition of the smoke varies greatly depending on the type of fire and the fuel source, but deposits can still be useful both in determining the original position of items and locating the seat of the fire. The visual methods so far have focused primarily on locating the seat of the fire. Having located the possible origin, studying the burn pattern for signs of ILR will indicate areas of interest for analysis. The burn pattern can be interpreted by the intensity and direction of burn. Fires burn in an inverse cone from the base up, and tracing back the burn pattern can possibly indicate a source. A hard edge to a burn, or pool burn, is classically indicative of a pool of burning liquid, namely an accelerant or ignitable liquid. Any area with hard edges should therefore be sampled for laboratory analysis. Analysis of the pool burn is essential, as it is possible that burning material from a higher source dripped onto the surface, mimicking the burn pattern of an ILR. An obvious sign of the potential presence of ILR is a strong petroleum or hydrocarbon odor. This might only be present with high levels of ILR but nonetheless is another key indicator that must be considered during investigation.

3.3 Accelerant Detection Canines (ADC)

Having located the seat of the fire, it is often essential to accurately locate the presence of any ILR. Volumes of accelerant residue may be as low as a few microliters or less, and unless the exact location of the residue is pinpointed, it is quite probable that the ILR will be overlooked when sampling. The use of *Canis lupus* var. *familiaris*, better known as the common dog, has been used by law enforcement and private agencies for the detection of many items of forensic interest including narcotics and controlled substances, accelerants and ignitable liquid residues, explosives, currency, cadaver and human remains recovery, and human scent tracking and lineup. The term

Figure 3.1 (see color insert following page 54) Accelerant Detection Canine Villain indicates the presence of lamp oil in five separate areas of a room with all validated by the lab, corroborating the statement of a dying victim. (Used with permission from Bill Whitstine, Florida Canine Academy, 19 Marshall Street, Safety Harbor, FL.)

"Accelerant Detection Canine" refers to a dog that is specifically trained to detect and indicate the presence of trace evidence in the form of ILR at a fire scene. While the dog is capable of detecting ILR, it is not the role of the canine to determine whether the cause of the fire was of malicious intent.

The dog is simply a powerful tool to assist the fire investigator. Figure 3.1 shows an ADC named Villain in action at an actual fire scene in which a woman dying from extensive burns stated that her boyfriend had chased her around the apartment pouring lamp oil on her and then ignited it. The boyfriend stated that she poured the lamp oil on herself and ignited it. Villain indicated five separate areas that contained lamp oil, with all areas confirmed by the forensic science laboratory. This evidence corroborated the girlfriend's dying statement and the boyfriend was convicted on first-degree arson and first-degree murder.

Likely, the world's first ADC was Mattie, who was trained in 1984 in the joint pioneering program of the Connecticut State Police (CSP) and the Bureau of Alcohol, Tobacco, and Firearms (ATF).[4] Mattie was first deployed by the Connecticut State Police in 1986, and by the late 1980s had been called to hundreds of fire scenes and demonstrated the value of ADCs as a valuable investigative tool. Canine detection of ILRs follows the same training framework of any canine detection, which includes imprinting of the odors that the canine is meant to detect, employing representative distracters or odors likely to be encountered in the field to which the canine should not be alert, and providing consistent reinforcement to ensure reliable results.

Commonly, the primary training odor presented to the dog is 50% evaporated gasoline; indeed, many handlers and trainers will testify that an ADC need only be trained on this one target to be alert to an ILR, but there is mixed opinion in peer-reviewed journals to either confirm or contradict this claim. Following the conditioning phase of training, it is essential that the dog be tested to ensure that there are no occurrences of false positives or false negatives. A false positive is defined as an alert when ILR is not present; a false negative is defined as no alert when ILR is present. Pyrolyzed debris commonly found at fire scenes is frequently introduced as a distraction medium. The canine must be capable of distinguishing the hydrocarbon aroma emitted by the pyrolysis of fire debris and true ILR. After the canine has been proven successful in the detection of accelerant odors without background distraction, the training focuses on the presence of ILR within fire debris samples. The dog is exposed to many fire debris and pyrolized samples, with only some of these samples positive for ILR. The dog is further trained until discrimination between ILR and pyrolysis products is achieved. Again, testing follows to ensure no false alerts. The culmination of training is an annual certification program that ensures accurate and effective training and subsequent reinforcement of the dog.

Unlike most electronic "sniffers" discussed in the next section, canines, when properly trained, can differentiate between accelerant residues and background pyrolysis products. Electronic "sniffers" generally give false positives from pyrolysis products such as those generated by burned plastics, foam-backed carpets, etc. Also, these mechanical sniffers must be brought to the source of the accelerant, which may be difficult to determine. Canines, on the other hand, can be trained to discriminate between accelerants and burned plastics, foam-backed carpeting, etc., and naturally scent to source (low to high concentration), pinpointing the highest concentration of accelerant residue available. After a canine alerts to a particular location, the fire debris from that area as well as a comparison sample away from this location is collected and placed inside a paint can, glass jar, or plastic bag with the former the most commonly employed and the latter also increasing in popularity.[5] Each container has drawbacks with the cans susceptible to corrosion from wet samples, glass jars susceptible to breakage, and plastic bags susceptible to punctures. After collection, the containers with the collected debris can be checked by the canines or electronic sniffers to ensure that adequate samples are collected.[6] Without a screening device at the scene, more samples must generally be collected; whether adequate samples were taken will not be determined until after laboratory analysis, at which point additional sampling is generally not possible.

More reliable instrumental methods including portable GC/MSs are under development and will likely be increasingly used at the scene to assist in the collection of reliable evidence. However, at present, canines still represent the state of the art in this regard and their agility and rapid scent-to-source capabilities are unlikely to be rivaled by machines in the foreseeable future. In a recent review by Yinon, he concluded that "electronic noses for detecting explosives are still in various stages of testing and have a long way to go before being field-operational."[7] In an extensive study by Tindall and Lothridge, 42 accelerant detection canine teams were tested to determine their capability to detect ILR from fire debris samples.[8] The teams were subjected to a variety of tests, including detection of ILR from pyrolysis debris, with a reported accuracy of 89.3% and 9 false positives and 28 misses from 130 positive samples, although multiple misses were reported as attributable to handler error, indicating a problem with the handler's training and confidence rather than the canine's ability. In a separate test in the same study, the dogs were exposed to different classes of accelerant including light, medium, and heavy petroleum distillates, isopar mixtures, and 50% evaporated gasoline, to compare the efficacy of training dogs solely on gasoline vs. a more broad-range training approach. While the dogs trained on various accelerants performed better than the dogs trained solely on gasoline, the difference was not sufficiently significant to raise serious concern over the gasoline-only approach. In a final test to determine detection levels, 80% of the canine teams achieved detection below reported instrumental limits.

Other studies have focused on detection levels and the selectivity/sensitivity of canine accelerant detection of ILR over background pyrolysis products. In one study by Kurz et al. featuring only two dogs, detection levels of kerosene, gasoline, and isopars were reported at 0.01 μl, equal to or better than a purge-and-trap headspace analysis by gas chromatography with flame ionization detection.[9] In a later study, 34 canines were put to the same test, which featured a greater emphasis on the presence of distracting pyrolysis odors. Detection levels were reported successful at 5.0 μl of gasoline, but the success of detection dropped to 50% or less for accelerant volumes of 0.05 to 0.2 μl.[10] Another observation of Kurz's second study was that detection of accelerants other than gasoline at the same levels was significantly less successful on dogs trained solely on gasoline samples. A wide range of variability was also observed among the handler–canine teams with most performing well on higher levels of gasoline but some having difficulty on trace levels of petroleum products in the presence of high levels of background, presumably due in part to less training with these conditions as many teams performed flawlessly throughout the tests. Tranthim-Fryer and DeHaan identified a variety of pyrolysis products from burned carpet and rubber underlays, which could contribute to false positives by canines.[11]

A commentary by Schultz, Ercoli, and Cerven highlights once again that some dogs are better than others at discriminating between accelerants and pyrolysis products, and points out that when small amounts of accelerants are present with large amounts of pyrolysis products, accurate laboratory identification is difficult using highly selective techniques, including MS detection.[12] Katz and Midkiff discuss the ongoing reliability issues of unconfirmed canine accelerant detection in court and conclude that until there is additional research providing improved understanding of how dogs react to target odors and the effects on this reaction by the presence of closely related odors, there will be continued controversy over the admissibility of canine testimony.[13] This issue actually is important for all types of detector dogs, and similar issues of admissibility in court have occurred for drug- and explosive-detection canines.[14,15] Unfortunately, until recently, there has been a limited scientific understanding of the complex process of olfaction, including canine olfaction. Emerging research is improving our understanding of the selectivity and sensitivity of canine olfaction, but many questions remain unanswered. Even the issue of sensitivity is often misrepresented in the popular media with canines attributed capabilities often untested scientifically. What is known is that canines possess larger olfactory tracts and bulbs compared to humans, with as many as 50 times more olfactory cells as seen in Figure 3.2. Canines also have a more efficient nasal anatomy with significantly more airflow passing directly over the olfactory receptors as compared to humans. The result is a superior sense of smell between canines and humans, but the levels of which are often tens to hundreds of times better rather than the millions to billions sometimes reported, and highly variable on the target compounds.

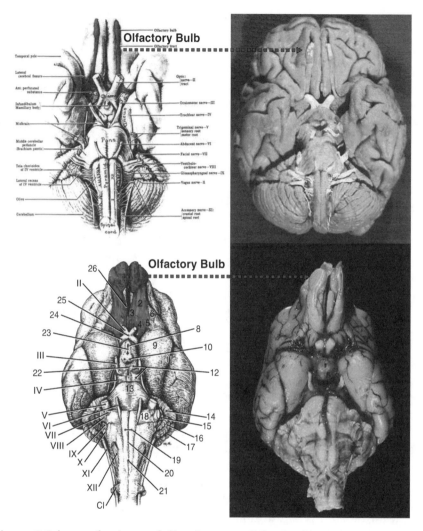

Figure 3.2 (see color insert following page 54) Comparison of human and canine brains, demonstrating the significantly larger olfactory track and bulb of the canine. (Sources: *Grant's Atlas of Anatomy*, Williams & Wilkins, Baltimore, 1978; http://www.sci.uidaho.edu/med532/cranialnervestartpage.htm; *Miller's Anatomy of the Dog*, 3rd ed., W.B. Saunders, Philadelphia, 1993; http://cal.vet. upenn.edu/neuro/server/lab4frameset.html.)

3.4 Instrumental Field Tests

The use of a portable field test such as a hydrocarbon "sniffer" is often used to complement visual and possibly canine indicators to locate any ILR traces. A hydrocarbon sniffer uses a small vacuum pump to pull vapor samples through a narrow nozzle directly into a chemical detector specifically designed to detect hydrocarbon and organic vapors. The technology applied to the chemical detection can take on several forms. One of the simplest chemical detectors for gas analysis is the paper tape system. A strip of paper tape is chemically impregnated with a reactant, which will change color when exposed to a specific target vapor. Paper tape is less prone to interference than other gas detection systems due to the specificity of the chemical reagent; however, the paper tape system is normally limited only to toxic gases. Paper tape technology cannot be readily applied to combustibles and as such is generally unsuitable for arson detection.

Electrochemical sensors are based on the reaction between a target gas molecule and an applied reagent to produce an electrical current that is measured by an instrument. Catalytic bead sensors use a combustion chamber to burn combustible gases on the surface of a catalyst bead. The resultant increase in resistance is then converted into concentration by the instrument computer. Electrochemical systems are inexpensive, quick to respond, and contain no moving parts, making them popular as portable devices. The sensors are sensitive but lacking in selectivity, such that false positives may be readily observed. The catalytic sensor requires recalibration after 3 months, and the typical lifetime of such an electrochemical sensor is only 2 years. Solid-state detectors employ a metal oxide transistor that changes resistance when the vapor molecule interacts with the transistor surface. This form of detector is inexpensive and boasts a considerably longer lifetime than paper tape or electrochemical sensing, but it is less selective and so gives rise to higher occurrences of false positives and erroneous alerts. Solid-state detectors often give a nonlinear response over concentration ranges, and in prolonged absence from exposure to the target molecules, the sensitivity of the instrument can deteriorate, increasing the difficulty of successful calibration.

One of the most effective forms of chemical detection and analysis of gas vapor is Fourier Transform Infrared Spectroscopy (FTIR), which studies the characteristic absorptions at different wavelengths of infrared light. While this technique is highly accurate with a low occurrence of false positives, the weight, cost, and complicated instrumentation of this detection method are prohibitive to the widespread use of FTIR as a portable field instrument.

One relatively inexpensive portable instrument (ca. $3000 with accessories) is the TLV Sniffer (Scott Instruments, Exton, PA)[16] shown in Figure 3.3. It features a catalytic bead chemical sensor, and like many field instruments it incorporates an audible alarm in addition to the needle gauge so that the user may concentrate on the fire scene without having to watch the needle on the meter. The advantage of using such a detector is its high sensitivity for hydrocarbon molecules. Recall however that the catalytic bead is a universal combustibles detector and, as such, false positive alerts on pyrolysis products and debris are highly likely due to their hydrocarbon nature. Nevertheless, this form of electronic sniffer is relatively inexpensive and, used in

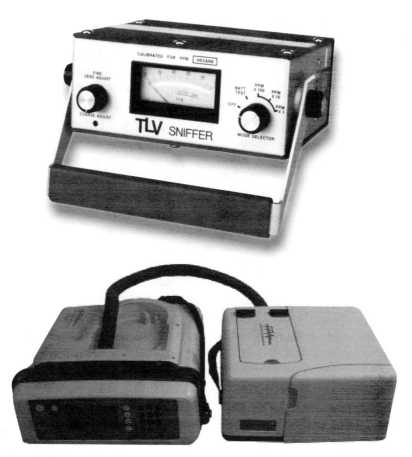

Figure 3.3 Examples of instruments available for field sampling of ignitable liquid residues at the crime scene. Top: Scott Instruments' TLV Sniffer. Bottom: Inficon's GC/MS.

combination with accelerant detection canines, can improve the efficiency of collecting quality evidence from the field.

A significantly more expensive electronic sniffer (ca. $60,000) which has been tested on accelerants from simulated fire debris is the AromaScanner Electronic Aroma Detection Device (AromaScan, Inc., Hollis, NH).[17] This system replicates a neural network of sensors more akin to the olfactory system of a mammal than the typical chemical sensors previously discussed. Rather than alerting to the detection of a single compound through a single chemical sensor, the aroma array applies a bank of sensors, each one different and designed to target specific components of the combined aroma. The neural network is then conditioned to target odors much like an accelerant detection canine. Early results have been promising for laboratory standards, but the application to fire debris has highlighted some problems with sample conditions. Factors such as vapor temperature and humidity have been observed to heavily influence the result obtained. Detection limits have not been established to this author's knowledge, but the levels at which the aroma detection system has been tested are up to 1000 times higher that those which a canine has been shown to reliably alert to.

Another relatively expensive portable instrument that has the potential to complete the normal lab analysis in the field is the gas chromatography/mass spectrometry portable GC/MS device. An example of such an instrument is the Inficon Hapsite® Field-Portable GC/MS (Inficon, East Syracuse, NY)[18] shown in Figure 3.3. The Hapsite® is a miniaturized quadrupole GC/MS weighing only 35 lb and specifically designed for the analysis of volatile compounds with a mass range of 1 to 300 amu. The interface between the GC and MS is a methyl silicone membrane that allows organics to migrate to the MS while venting most nonorganics and higher-molecular-weight compounds. The obvious advantage of the Hapsite® is the ability to get NIST-searchable MS spectra in the field to identify and semiquantitate unknowns. The disadvantages are the high cost (ca. $110,000) and the fact that run times are generally about 10 min each; however, it can also be operated in the MS mode only and operated as a real-time "sniffer" allowing for rapid screening of specific compounds. Instruments such as this will increasingly be used as their cost is reduced and effectiveness in the field is demonstrated.

Another approach employed over the years is to sample the ILR from a fire scene into adsorption tubes followed by instrumental analysis in the laboratory. This method allows for sampling over the same areas pinpointed by canines and/or electronic sniffers, yet minimizes the potential losses of ILRs associated with the collection, transportation, and storage of actual fire debris. Pioneering work in this area by Clausen demonstrated that systems employing portable air sampling pumps can efficiently trap ILR into silica gel adsorbent tubes either from over debris or even from active fire atmospheres with collectors carried in firefighters' coat pockets with subsequent thermal desorption of ILRs.[19] Fenkel and Tsaroom reported that field sampling using the adsorption tube method employing activated charcoal followed by solvent elution was effective for collecting ILRs in the field even in cases where sampling was normally unsuitable, such as open areas or rainy conditions.[20] A relatively inexpensive system (less than $2000) is commercially available, which allows for sampling in the field from debris using a heated Teflon chamber or difficult surfaces such as concrete into adsorption tubes for solvent or thermal desorption in the lab.[21]

One of the most recently employed methods in the lab, and increasingly in the field, for sampling ILRs is the technique of solid phase microextraction (SPME). SPME has been demonstrated to be a valuable concentration technique for the recovery of ILRs under a variety of conditions.[22,23] Examples of commercially available SPME holders are shown in Figure 3.4 and include field portable samplers available from Supelco, Inc. and Field Forensics, Inc.

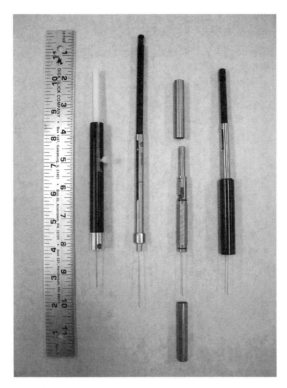

Figure 3.4 Examples of commercially available SPME holders. From left to right: Supelco portable sampler, Supelco autosampler, Field Forensics portable sampler, and Supelco manual sampler.

A more-detailed photograph of the Field Forensics assembly is shown in Figure 3.5 along with a close-up of the coated fused silica fiber as well as a collection of different fiber assemblies available from Supelco with color coding differentiating the different chemistries and either manual or autosampler modes. An SPME method has been developed which allows for the analysis of ILR from human skin or clothing, wherein a Kapac bag is wrapped around a subject's hand, for example, and the headspace is sampled by piercing the bag with the SPME fiber.[24] Field methods such as this may prove useful for confirming accelerant detection canine alerts to individuals, as it has been shown that arsonists often stay in the vicinity of their work and canines have been used to locate these individuals. It has been demonstrated that when pouring gasoline around a room, there is always a transfer of ILR onto clothing and shoes.[25] Figure 3.6 shows a representative chromatogram employing headspace SPME/GC/FID for recovering ILR components from a subject's hand 5 sec after contamination with diesel fuel. Figure 3.7 shows a representative chromatogram employing headspace SPME/GC/FID for recovering ILR components from a subject's hand 3.5 h after contamination with diesel fuel.

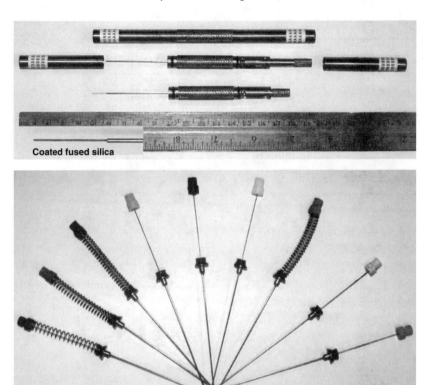

Coated fused silica

Figure 3.5 (see color insert following page 54) Top: Detail of Field Forensics' SPME holder. Bottom: Examples of different SPME fibers.

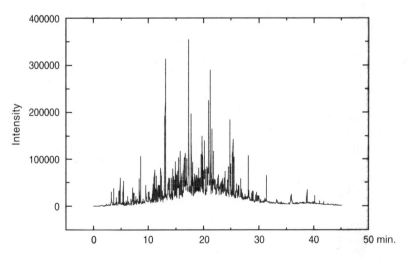

Figure 3.6 The gas chromatograph of a hand that was contaminated with diesel for 5 sec. The hand was then covered with a fire debris bag and SPME for 15 min with the first 5 min heating using a heat lamp.

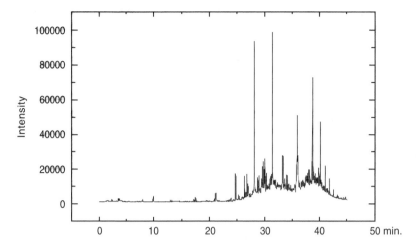

Figure 3.7 The gas chromatograph of a hand that was contaminated with diesel for 3.5 h. The hand was then covered with a fire debris bag and SPME for 15 min with the first 5 min heating using a heat lamp.

ILR can be sampled not only from the headspace of solid samples but also from aqueous samples or a combination of both coined partial headspace sampling where the SPME fiber is only partially immersed into an aqueous sample.[26,27] Table 3.1 illustrates the SPME/GC/FID area ratios for headspace, partial headspace, and direction immersion of ILR components, demonstrating that headspace SPME favors the recovery of the lower-molecular-weight components, with direct immersion favoring the higher-molecular-weight components and partial headspace providing a good recovery of the whole range. A considerable amount of water may be present in fire debris, and Figure 3.8 and Figure 3.9 demonstrate the effect of dry and wet conditions on the recovery of simulated ILR components using headspace SPME sampling, showing that in the presence of water a greater number of the higher-molecular-weight components are recovered (Figure 3.9).

3.5 Comparison of Techniques

Clearly, the visual inspection of a fire scene is a technique that should always be done regardless of the application of instrumental or canine detection or a combined approach. Table 3.2 summarizes a general comparison between instruments and canines. One of the biggest advantages of canine detection over the use of instruments is the dog's ability to track an odor to source; that is, the dog can follow a scent cone back to its origin, pinpointing the location of what it believes to be ILR or an area of interest. To accurately locate samples as small as a couple of microliters of ILR, the detection method

Table 3.1   Recoveries of Wet Accelerants[a]

Compound	Retention time (min)	Headspace (Area) Ratio	Partial Headspace	Direct Immersion
Pentane	1.37			
Hexane	1.70			
Heptanes	2.50		0.675	0.656
Toluene	3.18	(198) 1		
Octane	3.55	(3,381) 1	0.340	0.038
p-Xylene	4.70	(10,120) 1	0.318	0.010
Noncane	5.24	(27,273) 1	0.276	0.010
3-Ethyltoluene	6.88	(54,115) 1	0.266	0.007
2-Ethyltoluene	7.46	(63,312) 1	0.261	0.007
1,2,4-Trimethylbenzene	8.06	(71,978) 1	0.255	0.007
Decane	8.31	(98,600) 1	0.238	0.005
1,2,3-Trimethylbenzene	9.17	(73,905) 1	0.249	0.007
Butyl benzene	10.73	(160,478) 1	0.234	0.005
Undecane	12.73	(704,1230) 1	0.234	0.004
Naphthalene	16.13	(72,089) 1	0.215	0.010
Dodecane	17.02	(433,713) 1	0.278	0.005
1-Methylnaphthalene	20.52	(202,080) 1	0.215	0.008
Tridecane	20.94	(471,409) 1	0.399	0.012
2-Methylnaphthalene	21.08	(171,621) 1	0.216	0.010
Tetradecane	24.52	(392,384) 1	0.620	0.034
Pentadecane	27.80	(186,451) 1	0.893	0.112
Hexadecane	30.99	(78,509) 1	1.331	0.368
Heptadecane	35.2	(21,585) 1	2.335	1.120
Pristane	35.41	(31,102) 1	2.509	1.198
Octadecane	38.43	(6,118) 1	5.272	3.138
Nondecane	39.85	(1,469) 1	16.579	9.900
Eicosane	40.85	(394) 1	30.558	18.282
Heneicosane	41.63	(541) 1	30.053	16.911
Docosane	42.31		20.167	10.951
Tricosane	42.98		39.739	20.009

[a]   740 ml water was added on 1-μl accelerants in 1-liter can. Change (compared to headspace SPME) with PDMS fiber, different SPME sampling methods (SPME for 15 min) at room temperature.

must be very accurate to the ILR source. The use of an instrumental sniffer relies on the user to follow that scent cone by intuition and trial and error. So while the canine detection is based on a dog's natural ability to track a scent, the instrumental analysis requires a skilled user. Another significant advantage of canine detection over instrumental techniques is the dog's ability to discriminate between the background pyrolysis products and any ILR present. Where studies have shown that in most instances the dog will alert only in the presence of ILR, the hydrocarbon sniffer will generally have much greater difficulty in differentiating between the two vapors, depending on the instrument and conditions. This in turn will lead to more false positives

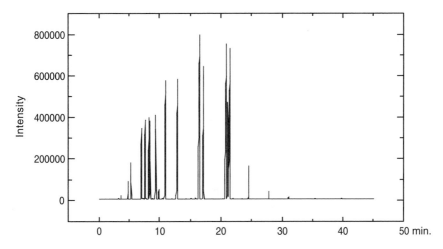

Figure 3.8 The gas chromatograph of dry accelerants by PDMS fiber headspace SPME 1-liter sample for 25 min.

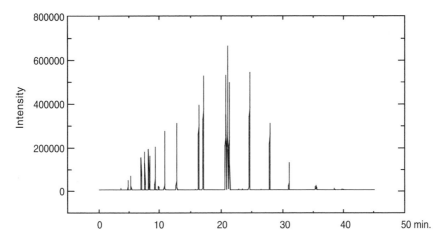

Figure 3.9 The gas chromatograph of wet accelerants (200 ml water was added on 1-liter sample in 1-liter can) by PDMS fiber headspace SPME for 25 min.

and thus a greater workload of samples submitted to the laboratory for analysis.

Areas in which the instrument outweighs the dog are operation time and the suitability of instruments for hazardous environments. In extreme conditions, a detection canine will often require a rest stop after 40 min of work, as the dog will tire quickly, whereas the instrument is limited only by the lifetime of batteries and these can be easily replaced. Studies have shown, however, that dogs can be trained to perform effectively for at least 90 to

Table 3.2   General Comparison between Instrumental Detection Devices and Detector Dogs for Ignitable Liquid Residues

Aspect	Instrument	Canine
Selectivity (vs. interferents)	Problematic	Very good
Mobility	Limited at present	Very versatile
Integrated sampling system	Problematic/inefficient	Highly efficient
Scent to source	Difficult with present technology	Natural and quick
Initial cost	Ca. $3,000 to 110,000	Ca. $6,000 to 7,000
Annual cost (excluding personnel)	Ca. $1,000 to 5,000 (service contract)	Ca. $1,000 to 2,000 (vet and food costs)
Duty cycle	Ca. 23 h/d (with multiple operators and minimal maintenance)	Ca. 8 h/d (longer periods possible with multiple handlers)
Calibration standards	Can be run simultaneously (i.e., chromatography based)	Sometimes run simultaneously but typically run individually
Operator/handler influence	Less of a factor	A potential factor
Environmental conditions	Less affected	May adversely affect (i.e., high temperatures)
Instrument lifetime	Generally ca. 10 years	Generally 6 to 8 years
State of scientific knowledge/ courtroom acceptance	Relatively mature/generally unchallenged	Late emerging/Sometimes challenged
Sensitivity	Very good	Very good
Speed of detection	Generally fast	Generally fast
Initial calibration	Generally performed by manufacturer	Generally performed by supplier
Operator training	Typically a 40 h course	Typically 40 h course minimum
Certifications	Typically annual	Typically annual
Recalibrations	Daily to weekly	Daily to weekly
Scientific foundations	Electronics, computer science, analytical chemistry	Neurophysiology, behavioral psychology, analytical chemistry
Potential effects on performance	Electronics/mechanical	Disease conditions

120 min of continuous searching but special conditioning and multiple handlers were performed in this study.[28] In hazardous conditions where it might be considered unsafe to take a canine, often an operator in full protective equipment may still use the instrument to search for ILR. Unlike a canine that only operates as part of a team with its dedicated handler, any person who is qualified and trained in the operation of a sniffer may use the instrument. Some canine detection programs do argue that any handler can operate any dog, but often a specific handler and canine will develop very strong ties and the canine will be reluctant to operate with another handler. Furthermore,

the replacement handler may not understand that specific dog's behavior such that the likelihood of false alerts or misses may be enhanced. This handler–canine team interaction and the variability in training and certification standards is one of the main perceived limitations of accelerant detection canines as there is considerable uniformity of instrumental detectors but significantly more variability in canine teams. Even with this variability, studies have shown that canines are more successful at separating ILR from pyrolysis products in the fire debris background, which makes them an essential component of a field sampling protocol in which reliable evidence is located (pinpointed), collected, and analyzed.

3.6   The Future

Regardless of the choice of method of fire scene analysis, it must be stressed that the confirmation or denial of arson cannot at present be determined at the fire scene, but only after forensic analysis of the samples collected at the scene. Either way, the use of field tests by instrument or canine is favored over simple visual analysis by many to reduce the number of samples collected, thus reducing the amount of work expected of the laboratory for each investigation. Electronic sniffers provide an additional confirmation of an ILR at the scene and are useful in employing a layered approach to provide the optimal presumptive identification and collection of ILR. Table 3.3 compares three different approaches to collecting and analyzing fire debris evidence, from the basic approach focusing on visual cues and lab analysis of debris to a layered approach, including combinations of biological and instrumental detectors and advanced collection methods and, finally, employing more advanced methodology including advanced laboratory techniques. Ultimately, identification of ILR on site should increasingly be a reality in the not-too-distant future. As instruments become more selective, more portable, and less expensive, instrumental methods, particularly GC/MS methods, will undoubtedly be increasingly used at the scene. Additional research and field testing will continue to improve the performance and consistency of accelerant detection canines and their use in the field should continue to expand. Canines' agility and ability to rapidly scent to source, as well as their ability to discriminate between ILR and pyrolysis products (with sufficient training), are unlikely to be rivaled by instruments in the foreseeable future. It is clear that the modern fire investigator must give thoughtful consideration to both instrumental (chemical) and canine (biological) detection techniques in combination with traditional visual cues and continue to stay abreast of field portable GC/MS instrument developments which could provide the on-site confirmation of ILR.

Table 3.3   Comparison of Different Approaches to Collecting and Analyzing Evidence[a]

	Basic approach with (hopefully) useful evidence collected	Layered approach with improved evidence collection	Advanced approach with improved field and lab analysis
Field analysis	Visually inspect and photograph scene	Visually inspect and photograph scene	Visually inspect and photograph scene
		Pinpoint useful samples with K9 and instruments	Pinpoint useful samples with K9 and instruments
		Adsorption tube collection of sample vapors at scene	SPME/GC/MS ID of samples/suspects in the field
		Interpret and collect additional samples as needed	Interpret and collect additional samples as needed
	Collect, preserve, transport samples to laboratory	Collect, preserve, transport samples to laboratory	Collect, preserve, transport samples to laboratory
Lab analysis	GC or GC/MS analysis, interpretation in laboratory	GC or GC/MS analysis, interpretation in laboratory	GC or GC/MS analysis, interpretation in laboratory
			Advanced analysis to improve sensitivity/selectivity
	Combine results, conclude testify in court	Combine results, conclude testify in court	Combine results, conclude testify in court

[a] From basic collection with laboratory analysis to a layered approach, including canines, machines, GC/MS analysis in the field, and advanced analysis in the laboratory.

References

1. Niamh Nic Daeid, University of Strathclyde Forensic Science Unit, personal communication.

2. DeHann, J.D., *Kirk's Fire Investigation*, 4th ed., Prentice-Hall, NJ, 1997.

3. Ide, R., *Crime Scene to Court: The Essentials of Forensic Science*, White, P.C., Ed., RSC Press, Cambridge, 1998.

4. Berluti, A.F., Sniffing through the ashes: Connecticut initiates canine accelerant detection, *Fire Arson Invest.*, 4, 31–35, 1989.

5. Kinard, W.D. and Midkiff, C.R., Arson evidence container evaluation II: New generation Kapac bags, *J. Forensic Sci.*, 36, 1714–1721, 1991.

6. Whitstine, W.H., Sniffing the Ashes/K-9s in the Fire Service, International Society of Fire Service Instructors, Aschland, MA, 1992.

7. Yinon, J., Detection of explosives by electronic noses, *Anal. Chem.*, March 1, 99A–105A, 2003.

8. Tindall, R. and Lothridge, K., An evaluation of 42 accelerant detection teams, *J. Forensic Sci.*, 40(4), 561–564, 1995.

9. Kurz, M.E., Billard, M., Rettig, M., Augustiniak, J., Lange, J., Larsen, M., Warrick, R., Mohns, T., Bora, R., Broadus, K., Hartke, G., Glover, B., Tankersley, D., and Marcouiller, J., Evaluation of canines for accelerant detection at fire scenes, *J. Forensic Sci.*, 39(6), 1528–1536, 1994.

10. Kurz, M.E., Schultz, S., Griffith, J., Broadus, K., Sparks, J., Dabdoub, G., and Brock, J., Effect of background interference on accelerant detection canines, *J. Forensic Sci.*, 41(5), 868–873, 1996.

11. Tranthim-Fryer, D.J. and De Haan, J.D., Canine accelerant detectors and problems with carpet pyrolysis products, *Sci. Justice*, 37, 39–46, 1997.

12. Schultz, B.A., Ercoli, J.M. and Cerven, J.F., Commentary on evaluation of canines for accelerant detection at fire scenes, *J. Forensic Sci.*, 39(6), 1528–1536 , 1994; *J. Forensic Sci.*, 41(2), 187, 1996.

13. Katz, S.R. and Midkiff, C.R., Unconfirmed canine accelerant detection: A reliability issue in court, *J. Forensic Sci.*, 43, 329–333 1998.

14. Furton, K.G., Hong, Y.C., Hsu, Y.L., Luo, T., Rose, S., and Walton, J., Identification of odor signature chemicals in cocaine using solid-phase microextraction/gas chromatography and detector dog response to isolated compounds spiked on U.S. paper currency, *J. Chromatogr. Sci.*, 40, 147–155, 2002.

15. Furton, K.G. and Myers, L.J., The scientific foundations and efficacy of the use of canines as chemical detectors for explosives, Invited review for special thematic issue: "Methods for Explosive Analysis and Detection," *Talanta*, 54(3) 487–500, 2001.

16. www.TLVSniffer.com, accessed on May 22, 2003.

17. Barshick, S., Analysis of accelerants and fire debris using aroma detection technology, *J. Forensic Sci.*, 43(2), 284–293, 1998.

18. www.inficon.com, accessed on June 12, 2003.

19. Clausen, C.A., Early detection and entrapment of accelerants in fire atmospheres, *Arson Anal. Newsl.*, 6, 105–140, 1983.

20. Fenkle, M. and Tsaroom, S., Field sampling of arsons area by the adsorption tube method, *Arson Anal. Newsl.*, 9, 33–41, 1986.

21. www.trilobyte.net/pas, accessed on June 6, 2003, and Johnston W.K., personal communication, Portable Arson Samplers, Tooele, UT.

22. Furton, K.G., Almirall, J.R., and Bruna, J., A simple, inexpensive, rapid, sensitive and solventless method for the recovery of accelerants from fire debris based on SPME, *J. High Resolut. Chromatogr.*, 18, 625–629, 1995.

23. Furton, K.G., Wang, J., Hsu, Y.-L., Walton, J., and Almirall, J.R., The use of solid-phase microextraction/gas chromatography in forensic analysis, *J. Chromatogr. Sci.*, 38, 297–306, 2000.

24. Almirall, J.R., Wang, J., Lothridge, K., and Furton, K.G., The detection and analysis of ignitable liquid residues extracted from human skin using SPME/GC, *J. Forensic. Sci.*, 45(2), 461–469, 2000.

25. Coulson, S.A. and Morgan-Smith, R.K., The transfer of petrol on to clothing and shoes while pouring petrol around a room, *Sci. Justice*, 112, 135–141, 2000.

26. Wang, J., Variable Influencing the Recovery of the Ignitable Liquid Residues from Simulated Fire Debris and Human Skin Using Solid Phase Microextraction/Gas Chromatography, M.S. thesis, Florida International University, 1998.

27. Furton, K.G., Almirall, J.R., Bi, M., Wang, J., and Wu, L., Application of solid-phase microextraction to the recovery of explosives and ignitable liquid residues from forensic specimens, *J. Chromatogr.*, 885, 419–432, 2000.

28. Garner, K.J., Busbee, L., Cornwell, P., Edmonds, J., Mullins, K., Rader, K., Johnston, J.M., and Williams, J.M., Duty Cycle of the Detection Dog: A Baseline Study, Final Report, Federal Aviation Administration, Washington, D.C., 2000.

Essential Tools for the Analytical Laboratory: Facilities, Equipment, and Standard Operating Procedures

4

CARL E. CHASTEEN

Contents

0-8493-7885-0/04/$0.00+$1.50
© 2004 by CRC Press LLC

4.1 Facilities and Equipment

4.1.1 Design

The great American architect Frank Lloyd Wright is quoted as having said, "Form follows function." Before designing a structure or workspace, there should be familiarity with the functions that are to be undertaken. The layout of the rooms, hallways, and storage spaces within a building should be intimately dependent on the work occurring in those areas. The flow of personnel through the facility as they utilize distinct workspaces must be considered in the design. Tasks are not necessarily confined to one area of a building. For a forensic laboratory, this axiom is especially true. The designer must realize that there will be a semi-public area near the entrance to the laboratory. A secure evidence intake area with an even more secure evidence storage area must be planned. Security must also be considered in the design of the evidence preparation and testing areas. The workspace, where analysts will review information and create reports, must be functional as well as comfortable for the employee. Knowledge of the key factors regarding the techniques, equipment, and instrumentation is essential for design of the workspace.

    With limited budgets, few new forensic laboratories are being built. Most laboratories are situated in existing facilities and must adapt or remodel available workspace to accommodate changes in employees, techniques, and instrumentation. As a resource for those who are designing or remodeling a forensic laboratory, the National Institute of Justice issued a 1998 research report titled "Forensic Laboratories: Handbook for Facility Planning, Design, Construction, and Moving."[1] This report provides a general guide for many of the "function" aspects of a forensic laboratory and translates them into the "form" arena.

"Forensic Laboratories: Handbook for Facility Planning, Design, Construction, and Moving" provides an excellent general resource, but it does not provide significant current information on design requirements regarding fire debris analysis. The report contains only a three-line section on the design of an Arson Investigation Room.[1] It refers to space for vacuum extraction. It is unclear to which preparation method they refer. In a survey conducted by forensic laboratories performing fire debris analysis,[2] the Technical Working Group for Fire and Explosions (TWGFEX) found that none of the 216 survey respondents used vacuum extraction. The term "vacuum extraction," may have been referring to the dynamic headspace technique where a vacuum is sometimes used to pull vapors through an adsorption tube. If so, they have not fully discussed the design needs for the technique. The design requirements for other fire debris evidence preparation techniques are not addressed. I refer to the techniques currently accepted by the majority of fire debris analysts and described by the American Society for Testing and Materials (ASTM).[3] Considering the breadth of materials covered in the report, it is unlikely that a full exposition of all of the necessary space for equipment or instrumentation for every forensic discipline could be included. This section will discuss key "function" issues that will aid in determining the "form" necessary to meet current practices in fire debris analysis.

4.1.2 Key Factors

There are numerous factors important in the design of a forensic laboratory. The heating, ventilation, and air conditioning (HVAC) system is essential to the comfort and productivity of the people working in the facility. Proper environmental controls are essential, as most instruments and computers must operate within a limited range of temperature and humidity. Additionally, factors such as ergonomics, ingress, and egress should be important considerations in the design of any workspace. Most of these are beyond the scope of this short chapter, but are covered in the previously referenced "Forensic Laboratories: Handbook for Facility Planning, Design, Construction, and Moving." The focus here will be on the key technical and personnel issues relevant to fire debris analysis.

4.1.3 Technical

Technical factors focus on the requirements for the various extraction techniques, the equipment for the extractions, the instruments to be used, the hazardous materials to be encountered, and the flow of the evidence. The design aspects of the laboratory must relate to the function of how the analysis is to be completed, what tools are required, and where they should be located.

4.1.4 Discipline

Fire debris analysis may utilize a variety of techniques for the extraction of ignitable liquids from the submitted evidence. Each has differences in the layout requirements for the workspace. The area for extraction of debris and handling of standards may be independent of the instrumentation area. If preparation and instrumentation are to be placed in the same area, the space requirements for the area are increased.

Distillation methods will require space for the distillation apparatus as well as areas for the cleaning and drying of glassware. The solvent wash method will require significant quantities of glassware to be at the ready and will affect how much storage space is to be incorporated into the design. While several of the techniques require the use of pure solvents, the solvent wash technique will require significantly greater quantities. This will affect bench storage requirements. For safety purposes, the bench storage area will need to be designed for storing ignitable liquids. The headspace techniques will require the use of ovens or hotplates for the heating of the samples. In order to mitigate the effects of vapors from evidence and solvents, all techniques will require a fume hood. Lastly, dedicated areas sufficient to store the minor pieces of equipment in a manner free from cross-contamination must be included in the preparation area.

Fire debris analysis has been placed as a subdiscipline of the trace evidence section in many forensic laboratories. Only a few laboratories have a dedicated fire debris analysis section. Only three states have laboratories where fire debris analysis is the primary discipline. These are found under the State Fire Marshals in Ohio, Texas, and Florida. This indicates that most public laboratories working fire debris analysis must share space and equipment with the other disciplines in the trace section.

Early fire debris analysts were expected to open the container of debris and sift through it seeking to identify items that could be relevant to the fire. They would look for the items of debris that indicated a pour pattern or trailer and isolate them for steam/vacuum distillation or solvent wash. The analysts would also examine the debris for other types of trace evidence and isolate them for further testing. These would include classic "trace" items, especially the remains of devices. The evidence could be opened and exposed to evaporation of the trapped ignitable liquids as the distillation techniques were effective only if there were relatively large quantities of ignitable liquid remaining in the debris.

For those performing the solvent wash method, a speedy examination was required. Solvent wash can be performed either on the entire submitted sample or on a selected portion of the debris. It is more effective in recovering trace levels of ignitable liquid than the distillation methods, but the process of preparing the evidence should not allow exposure of the debris to evaporation for any significant period. Unless the entire container of evidence was to be washed, the analyst would need to expose the debris in order to select their sample.

Conventional wisdom over the years has changed the thinking about fire debris and the investigation of fires. In most jurisdictions, the scene investigator has the responsibility of examining the debris for items of interest to other forensic disciplines. It is rare that the analyst in the laboratory will be called on to open and sift through the debris from a fire. The analyst's focus should be on finding and identifying any trace ignitable liquids trapped in the debris. Recovery methods and instruments have reached a level of sensitivity where trace levels of ignitable liquids can be found. Sometimes ignitable liquids are found that are inherent to the matrix.[4] Such increased sensitivity increases the potential for cross-contamination of the samples in the laboratory unless sufficient precautions are taken.

When looking for such trace quantities of ignitable liquid, the unprotected exposure of the debris to ambient conditions should be minimized. The debris should not be opened and spread across the bench. To do so increases the potential for evaporation and loss of trapped ignitable liquid. Any time an evidence container is opened, there is a potential for contamination. The time that an evidence container is to be open and exposed to foreign vapors must be minimized.

Fire debris evidence typically contains burned materials. Often, the materials, or portions of them, have been reduced to carbon. The components of ignitable liquids readily adsorb to carbon. Exposure of a can of debris to an atmosphere laden with ignitable liquid vapors raises the potential that those vapors may adhere to the debris. Thus, the design of the preparation area must include a fume hood that provides a constant airflow with sufficient internal workspace for the processing of the evidence containers.

The need to spread the debris for sifting will be limited to the occasion when the analyst needs to assess the container's specific contents. This will most likely follow the extraction procedure. The analyst will need sufficient bench space for examination to determine the condition of the evidence and to verify if the information on the evidence container matches the information in the submission documents.

A review of the data regarding methods and instrumentation from the Collaborative Testing Service,[5] the International Forensic Research Institute,[6] and the Technical Working Group on Fire and Explosives,[2] clearly shows that the majority of forensic laboratories utilize an adsorption/elution extraction technique employing charcoal/activated carbon strips as the adsorbent. The same data indicate that the majority of the responding laboratories use gas chromatography (most with mass detectors) as the instrument of choice for analysis. With these factors in mind and noting them as a growing trend in

laboratories, the remainder of this section will focus on the design requirements for dynamic and passive headspace extraction methods and gas chromatography as an instrumental technique.

4.1.5   Equipment

Both the passive and dynamic techniques require a fume hood with sufficient
internal space to allow an evidence container to be opened. This is necessary
to evaluate the contents and allow for placement and/or removal of the
carbon strip or adsorption tubes. The space in the hood should be sufficient
to hold the materials required for introduction of an internal standard into
the debris (if one is used). Hood space will also be required at the end of the
extraction process for placing carbon membranes into vials. Both the passive
and dynamic methods will require space for either addition of solvent to the
vial or elution of a solvent through the adsorption tubes.

For the passive technique, laboratory ovens should be used. The temperature can be regulated and will be more stable. The ovens may be placed
inside the fume hood. This will increase the size requirement of the fume
hood. The ovens may also be placed on the bench next to the fume hood,
provided the hood generates sufficient negative airflow in the room to remove
vapors. The ovens should only be operated after the activated carbon mem-
branes/strips have been placed into the evidence containers to be tested.
Likewise, the ovens need to cool and be vented before evidence containers
are opened for removal of the carbon membranes.

The dynamic headspace method requires that the evidence container be
fitted with both an input tube and an exit tube. Ovens may be used for the
dynamic technique but would have to be fitted to allow the connection of
either inert gas lines or vacuum lines to the tubes inserted into the evidence
containers. The dynamic technique is more efficiently performed when the
evidence containers are heated on a hotplate or in a mantle. If more than
one container at a time is to be extracted, multiple hot plates/mantles will
be required and this will increase the design requirement for bench and hood
space. A fume hood equipped with either a vacuum output or inert gas input
would be ideal for the dynamic technique. This requires the incorporation
of tubing and valves into the design of the fume hood.

4.1.6   Instrumentation

Gas chromatographs have a footprint that typically fits into a space 0.75 ×
0.75 m. Gas chromatographs which have mass detectors attached will require
more bench space both to accommodate the detector and vacuum pump. A
gas chromatograph will require a bench with sufficient strength to support
its weight. The bench should be stable if the vacuum pump of the MS is to

be kept on the bench rather than the laboratory floor. The vibrations of the pump could cause problems with the electronics of the instrument.

The gas chromatograph will require a supply of gases for operation. For a gas chromatograph with mass spectrometer, a minimum of one gas, for the carrier, will be needed. If chemical ionization is to be used, the instrument will require additional gas supplies or solvent inlets. For instruments with flame ionization detectors, the carrier gas and an additional supply of air/oxygen and hydrogen will be needed. All of these gases should be piped to the instrument area from a secure source (compressed gas cylinders, air compressor, or gas generators). Some areas have fire or safety codes that require flammable gases to be located either in a remote area or outside the laboratory. Plumbing of gas lines requires a significant amount of planning in order to minimize the amount of tubing for the gas lines and the placement of valves in the lab. The longer the path for the gas the more the pressure of the gas will be reduced by the time it reaches the instrument. This can be especially critical with a carrier gas. Fluctuations in gas flow can seriously affect the quality of the chromatography. Multiple instruments using the same source of gas should have sufficient pressure coming to them that they will not be affected if one instrument goes offline. All gases should be appropriately filtered to remove contaminants and ensure a clean baseline. This will have to be incorporated into the plumbing design as well.

Gas chromatographs and their peripheral devices (data stations, printers, vacuum pumps, etc.) require electricity. The design of the space where a gas chromatograph is to operate must consider the placement, voltage, and source of the electrical supply. The chromatograph and peripherals will require a sufficient number of outlets. Most will require normal household current of 120 V, and some instruments will require at least one 220-V outlet. Additionally, these should be dedicated circuits to minimize the effects of multiple power draws and fluctuations. To negate the effects of power fades, spikes, or outages, the installation of an uninterruptible power supply (UPS) should be considered. This may be a single-source UPS dedicated to a specific instrument, or a laboratory-wide UPS. Laboratory UPS units that can supply clean filtered power are available in configurations supplying 25 kVA and higher. With these units, the outlets for the instruments and other critical hardware are wired directly to the UPS. UPS units can be connected to emergency power generators. Such a configuration would assure the laboratory of continuous power in almost any situation.

Gas chromatographs are constantly venting carrier gases. In addition, capillary gas chromatographs operating in the split mode will be venting trace quantities of the solvent and analyte. Vacuum pumps produce significant quantities of vapors as well. The venting of these gases and vapors from the lab must be considered in the laboratory design. The laboratory should be designed with sufficient negative airflow to remove these vapors from the work area. Not only is this a necessary consideration for the employee's well being, but also eliminates the potential for contamination of evidence. Vacuum pumps used by gas chromatographs with mass detectors produce noise. This, as well as the above gas and effluent issue, indicates that instrumentation should be isolated to minimize their effect on laboratory personnel.

No laboratory design would be complete without the consideration of network cabling. The data station for the instrument should not be the only place in the laboratory where the data can be viewed and interpreted. With the use of a local area network and shared drives, instrument software can be placed on any computer in the laboratory. Laboratory information management systems (LIMS) and bar coding to track evidence can be linked to the instruments as well. With a comprehensive design, analysts can initialize samples in autosamplers and process the collected data from the comfort of their individual workspaces. In planning the laboratory, network outlets should be placed in the instrument locations as well as any location in the laboratory where data is to be entered into the LIMS. Cables meeting the latest networking requirements (currently enhanced category five) should be run from each outlet to the primary server in the laboratory. Space for the network hub should be planned in laboratory design as well. Here you will need a clean, cool space where the servers, routers, and network switchboards can be isolated.

4.1.7 Hazardous Materials

A variety of materials are submitted for fire debris analysis. The largest group of materials submitted to the laboratory for analysis are burned debris and construction materials. A review of the submission statistics for the Florida Fire Marshal's Laboratory from January 1993 to January 2003 was made. It indicated that, of the 29,055 samples submitted, 15,854 samples, or 54.6%, were categorized as either burned debris or a building material (carpet, flooring, concrete, wood, etc.). The remaining samples included tissue samples, cloth/clothing, paper, liquids, and miscellaneous items. Most samples will contain only trace quantities of ignitable liquids or other chemicals deemed as hazardous. In some instances, the samples may contain a more significant quantity of ignitable liquid. On occasion, the laboratory will receive materials that contain biohazards. These may be tissue samples from bodies found at the scene or clothing from suspects or victims.

The value of both the passive and dynamic headspace techniques is that they are relatively noninvasive of samples. Normal workplace controls and the use of personal protective equipment (latex gloves, eye protection, laboratory coats) should be sufficient to protect the employee from exposure to biohazards or other hazardous chemicals. So long as samples are opened only inside a functioning fume hood, any contamination inside the laboratory will be minimized. The passive and dynamic techniques isolate the samples within their submission containers and allow them to be resealed after extraction. Other techniques (solvent wash and distillation) require a facility plan for disposition of materials. This plan should include a materials isolation system within the area where extraction occurred. This may include utilization of bench space within the fume hood so that solvent-laden materials can evaporate off any residual solvent. Once the debris is sufficiently clean of solvent, it may be safely disposed. Items containing biohazards will need to be disposed of in a manner complying with the federal Occupational Safety and Health Administration (OSHA) standards.

Also of concern is the issue of disposition of liquid quantities of solvent in the milliliter or higher amounts. The facility may have a designated collection tank for these solvents. Only the same types of solvents should be poured together. Care must be taken to prevent the reaction of liquids when mixed in the same container. There are hypergolic mixtures that will react violently if mixed. An example would be a polyester resin and methyl ethyl ketone peroxide. The reaction is not immediate, but experiments at the Florida Fire Marshal's forensic laboratory have shown that the reaction will produce a significant fireball. Most liquids collected after the processing of evidence will be solvents and petroleum distillates that will be fully miscible. Once collected, the full container of material should be disposed of in a manner meeting both environmental and occupational requirements.

All laboratories will accumulate reagents, solvents, and ignitable liquid standards. All are to varying degrees considered hazardous. While limited quantities of these materials may be safely stored in the preparation areas, quantities of a liter, kilogram, or more should be stored in a hazardous materials room. This will be a room equipped with ventilation to draw vapors from the room and prevent their collection. The room should also be designed to be free of open electrical outlets. Explosion-proof lighting fixtures are recommended. The room should be positioned so that one wall is the exterior laboratory wall. Common sense dictates that this room should not be located in or near employee workstations or electrical equipment. Open shelving units are not adequate for the storage of hazardous materials. Specialty storage cabinets for flammables, corrosives, and poisons can be purchased from safety supply dealers.

Clearly, the most important design item in preventing the release of hazardous materials is planning. The second most important item of equipment is a fume hood. Fume hoods must be placed in the areas where the potential for exposure to hazardous vapors are highest. They should also be placed so that they will not impede the egress of employees should an accident occur. Safety showers should be installed in any laboratory section where chemicals will be handled. Fire blankets and fire extinguishers should be located near any extraction site. A wet sprinkler system is recommended for any laboratory. Incorporation of a sprinkler system into the design of a new facility is much less expensive than retrofitting a system into an existing laboratory.

4.1.8   Evidence Flow and Security

In the design of the laboratory, a secure dedicated area for evidence intake is required. As most samples will need to be stored for a short period before extraction can begin, the evidence intake area needs to be located as part of, or adjacent to, an evidence storage area. A secure area has access limited to only those employees with sufficient clearance for contact with evidence. This access can be limited by simply issuing keys for specific doors to select employees. This can prove to be a rather bulky option as the number of keys issued to a single individual may become significant. Programmable magnetic door locks with keypads, card readers, or biometric access pads (thumbprint readers) are the high-technology options. Their value is that a monthly report can be obtained showing the identity and date/timestamp of every person who accessed a particular door.

A requirement in most forensic laboratories is that evidence should be internally tracked. A history of the movement of the evidence from one station to another provides evidence of the secure handling of the evidence. This evidence "tracking" can be done either on paper or via electronic input to a Laboratory Information Management System (LIMS). A LIMS is essentially a cross-relational database and can be purchased as a packaged product with little variability or as a customized application. They are used to store the data concerning a sample's input. The data can be linked to the tracking and movement of the sample and allow the analyst to input descriptions of the evidence and results. Some may be linked to the instruments so that data is directly input. LIMS can be configured to pull data from various sources and compose the final laboratory report. They have become an essential tool for managing case data and providing statistical analyses.

One of the marvels of current technology is the use of barcodes to track evidence. The evidence container, case file, and sample extract vials can have a barcode applied at the time of evidence intake. When the evidence is to be moved, it is scanned and a notation is sent to the LIMS that it is leaving a specific area. The use of the LIMS can also record the identity of the employee making the move. The same evidence is again scanned when it is moved to the preparation area. The bar-coding system will be able to track the evidence extract as it is moved to the analysis area. If bar-coding is to be a part of your laboratory's design, network connections in all the rooms where the evidence will be moved will be essential. If a barcode system is unavailable, paper records indicating the internal laboratory movement of evidence should be utilized. This allows the analyst to clearly show the laboratory chain of custody.

As stated earlier, the evidence storage area should be located in close proximity to the area where the evidence will first be processed. This ensures that there is little risk of loss or misplacement of the evidence. Likewise, the area where the extracts are prepared should be in close proximity to the analysis area where they will be placed into the instrument. The instruments also need to be in a controlled access area. Of what value is controlled access to the evidence storage and preparation if the analysis is not secure as well?

4.1.9 Personnel

The design of the laboratory must consider the needs of the personnel who are to work there. Earlier discussion focused on the need for proper environmental controls for employee comfort. Personnel conducting the analyses must also have a sufficient and comfortable work area where they can concentrate on the data collected from the instruments. The use of networked software with shared access to the data collected by the instrument will avoid bottlenecks at the instrument and allow analysts to make their determinations from their workspaces in other areas of the laboratory.

4.1.10 Safety

The safety of personnel should be of paramount concern. Personal protective equipment and training in using it is critical. One procedure that has the dual purpose of protecting the evidence as well as the personnel is the insistence that evidence should only be opened inside a fume hood. Only one item of evidence opened at a time will also reduce the potential for mishap. A full discussion of ergonomic design is beyond the scope of this chapter. Ergonomics considers the design of the workspace, furnishings, and equipment in light of employee safety and comfort. One item of ergonomic importance that should be incorporated into the design is the placement of antifatigue floor mats near the fume hoods or benches. These are areas where employees will be standing for relatively long periods of time. These will help the employees to be able to work longer and more safely.

4.1.11 Adaptability

Laboratories must plan for the obsolescence of equipment and techniques. Over the past 20 years the methods and instruments of choice for fire debris analysis have changed. The next 20 years will produce even more changes. In order to address this eventual obsolescence, the design of the laboratory must include features that will allow it to adapt.

Bench space should be planned so that it can accommodate the size and weight of new instruments. This means that the benches should be larger than currently used and sufficiently braced to hold double the weight of current instruments. Raceways with simple access panels installed in the walls or ceilings of the laboratory will allow for changes to the plumbing of gases for instruments. Multiple dedicated electrical lines for both 120 and 240 V outlets should be planned at adequate intervals along the bench in the instrument area. The incorporation of modular laboratory furniture into the laboratory design will allow greater flexibility in adapting the workspace to changing requirements.

4.2 Sample Analysis

4.2.1 Equipment

The equipment to be used in fire debris analysis is dependent on the extraction methods and instruments chosen. The reports of two of the national proficiency testing organizations clearly indicate that the passive headspace method found in ASTM E1412, "Standard Practice for Separation of Ignitable Liquid Residues from Fire Debris Samples by Passive Headspace Concentration with Activated Charcoal" is the most common method for fire debris preparation. The same reports indicate that the use of gas chromatography with mass spectral detection, as described in ASTM E 1618 "Standard Test Method for Ignitable Liquid Residues in Extracts from Fire Debris Samples by Gas Chromatography/Mass Spectrometry," is the most common analytical method used. This section will focus on these techniques as well as alternatives.

4.2.2 Sample Preparation

Fire debris analysis requires that trace ignitable liquids be isolated from the debris. Once isolated, the extract must be in a form that can be instrumentally analyzed. The choice of technique for preparation of the sample has changed over time as new discoveries and methods are developed and validated. The choice of technique should factor the sensitivity of the recovery with time needed to perform an efficient extraction. Additional factors to consider in the selection of an extraction technique include the technique's effect on the evidence. A technique that substantially alters or destroys the evidence will negatively affect the ability of other analysts in retesting the evidence. Reexamination of evidence is sometimes necessary when results of initial laboratory analyses become disputed. For this reason, the analyst should use a technique that will either preserve the evidence for reanalysis or, at a minimum, preserve and archive the extract. The passive headspace method of ASTM E-1412 meets this requirement.

The technique chosen for sample preparation must be sufficient to recover the full range of components from the debris associated with the recognized classes of ignitable liquids. Most ignitable liquids are mixtures of aliphatic and aromatic compounds and may cover a range of molecules from pentane (C5) through normal eicosane (C20) and higher. Ignitable liquids may contain aromatic compounds with a wide range of complexity and containing various functional groups. Common aromatic compounds include benzene and toluene or condensed ring aromatics such as naphthalene. For the identification of gasoline, the recovery of alkylbenzenes is critical. It is the presence or absence of these compounds and the ratios they have with each other that allow the analyst to differentiate between ignitable liquids. For that reason, the technique chosen must be sufficient to recover multiple organic compound classes. Alcohols and ketones are rarely found in fire debris analysis, but the preparation technique should be sufficient to recover them. Ideally, the compounds and their ratios to each other in the profile of the extracted ignitable liquid should approximate the profile of a neat or diluted liquid standard.

4.2.3 Techniques

Different recovery techniques have been described over the years and can be found in both the forensic and analytical literature. The American Society for Testing and Materials E-30 Committee on Forensic Sciences has consolidated and refined the primary techniques and describes them fully. As of the writing of this chapter, six ignitable liquid recovery (extraction) techniques are described by ASTM:[3]

- E1385: Standard Practice for Separation and Concentration of Ignitable Liquid Residues from Fire Debris Samples by Steam Distillation
- E1386: Standard Practice for Separation and Concentration of Ignitable Liquid Residues from Fire Debris Samples by Solvent Extraction
- E1388: Standard Practice for Sampling of Headspace Vapors from Fire Debris Samples
- E1412: Standard Practice for Separation of Ignitable Liquid Residues from Fire Debris Samples by Passive Headspace Concentration with Activated Charcoal
- E1413: Standard Practice for Separation and Concentration of Ignitable Liquid Residues from Fire Debris Samples by Dynamic Headspace Concentration
- E2154: Standard Practice for Separation and Concentration of Ignitable Liquid Residues from Fire Debris Samples by Passive Headspace Concentration with Solid Phase Microextraction (SPME)

Over the years, other techniques have been described in forensic and analytical literature. Most are essentially variations of the techniques described above. An example would be the use of vacuum distillation as opposed to steam distillation. The use of methods employed by the Environmental Protection Agency for identification of ignitable liquids should not be applied to fire debris samples, as they do not take the contribution of pyrolysis or materials inherent to fire debris matrices into account. The ASTM techniques have different requirements. In the following sections, the techniques that utilize the equipment described will be identified in parentheses.

4.2.4 Fume Hoods (All Methods)

As discussed earlier in this chapter, fume hoods are essential for all the methods used in fire debris analysis. Some may argue that a fume hood is of minimal importance. Safety and contamination issues are of sufficient importance that even the simplest procedure should be conducted inside the fume hood. OSHA regulates the use of chemicals in the laboratory in their "*Occupational Exposure to Hazardous Chemicals, 29 CFR 1910.1450.*"[7] The regulation requires all laboratories to have a chemical hygiene program. Fume hoods are an integral part of a chemical hygiene program as they prevent exposures by removing vapors, gases, and particulates from the work area.

There are various designs and options available for fume hoods. Options include features such as multilevel sashes with sliding doors and automatic baffles that adjust to maintain a constant velocity of airflow regardless of the size of the opening of the sash. One option is the use of continuous flow fume hoods. These fume hoods, which cannot be turned off by employees, provide for constant venting of fumes. This eliminates the build-up of fumes that may present a potential danger. If continuous flow fume hoods are installed, they will have an impact on the HVAC system of the laboratory. Not only will they be venting fumes, but also they will be removing the conditioned air of the laboratory. It will be necessary to have the HVAC system rebalanced after installation. To attempt to mitigate that problem a "reduced flow fume hood" may be considered. These systems operate with reduced exhaust volumes. They are often less expensive than conventional hoods and can provide savings in operational costs.

Both ducted (removal of vapors completely from the facility) and ductless (passing vapors through a filter with the air returning to the room) are available. For fire debris analysis, a ducted fume hood with a filter that will capture organic vapors before venting the air to the atmosphere is preferred. This removes harmful vapors from entering the environment. The regular replacement of the filters would need to be incorporated in the laboratory's standard operating procedures to ensure that they did not become a potential source of contamination.

A fume hood should contain adequate space for the procedure to be conducted without requiring analysts to place more than their hands and arms inside the hood. Space must also be available to accommodate sharps and broken glass containers inside the hood. This eliminates their unprotected transport across the laboratory. In some laboratories the oven, hot plate, or heating mantle used in extraction methods will also be place inside the hood. For E1413, an inert gas or vacuum line plumbed directly into the fume hood may be advisable. If an oven, hot plate, or heating mantle is to be placed into a hood, the hood will also require electrical outlets. As the placement of electrical outlets directly inside the hood may be seen as a potential ignition source for ignitable or explosive vapors, they should be located outside the hood.

The face velocity of a fume hood is often used as the measure of the hood's performance. Face velocity is the measure of the average velocity at which air is drawn from the face to the exhaust of the hood. The National Fire Protection Association (NFPA) recommends a face velocity of 80 to 100 ft per min.[8]

4.2.5   Heat Sources (E1385, E1388, E1412, E1413, and E2154)

For the above methods, a mechanism for heating the sample is usually required. While all of the techniques, other than steam distillation, can be accomplished without external heating, optimum and recommended performance of the techniques will require heat to vaporize ignitable liquids trapped in the debris. For steam distillation, a heating mantle that conforms to the exterior contours of the extraction flask is ideal for distributing the heat. This will improve the efficiency of the distillation. The use of open flames as a heat source should be avoided due to the nature of the ignitable liquids. This is true for the steam distillation method, and the adsorption–elution method as well. While they may not have a large build-up of ignitable liquid vapors, the solvents used for eluting adsorbents are usually highly flammable.

A heating mantle or hotplate is often used in E1388, simple headspace. It is usually sufficient to raise the temperature to a point where volatiles will be driven into the headspace of the container so that a vapor syringe can be used to remove a sample for injection. For E1412, passive headspace, a more controlled source of heat is required. Allowing the temperature to rise too high for too long a period, or heating that is uneven, will affect the method's efficiency in concentrating ignitable liquids on the adsorbent;[9] for this method, a laboratory oven is advised. It will allow for a more precise control of the temperature. An automatic shutoff timer can be attached to the oven to control the heating time.

Internal experiments at the Florida Fire Marshal's laboratory have shown that the use of a vented oven (where the heated air is drawn through and out of the cabinet) maintains a more constant temperature on the debris samples than conventional static ovens. Additionally, the vented oven aids in the removal of incidental odors and vapors that escape from poorly sealed evidence containers. For E1413, dynamic headspace, controlled heating of the sample is also required. However, the presence of extraction tubes inserted through the exterior of the evidence container and into its vapor space may not allow the sample to be placed in an oven. Instead, a precision controlled heating mantle (that fits the evidence container) may be preferred. The key is that the sample must have a regulated and controlled amount of heat applied.

4.2.6   Adsorbents

The 1998 survey of fire and explosion analysis by the Technical Working Group for Fire and Explosives found that 79.8% of the respondents used the ASTM standards when performing fire debris analysis. Of that group, the passive headspace technique described in E1412 was the most commonly relied-upon method for extraction.[2] The original work in passive headspace utilized activated charcoal.[10] Currently, the adsorbent most often used in this method is the activated carbon strip or carbon membrane. The passive head-space technique has proven to be a simple but sensitive technique that has the additional advantages of being nondestructive to the sample and easily archived. Yet use of this adsorbent has several factors that must be carefully controlled in order to be effective. Newman, Dietz, and Lothridge determined that the optimal membrane size is 100 mm^2.[9] Additionally, the time for extraction and temperature must be carefully controlled.

The carbon membrane has a finite number of active sites available. The active sites of the membrane prefer large molecules that are typically less volatile. Because these take longer to volatilize, the strip has a tendency to be filled first by the smaller more volatile molecules. As the strip continues to be exposed to the vapors from the debris, the larger molecules will begin to displace the smaller molecules. If the temperature of the debris or the time for extraction for the debris extends beyond the optimum, the heavier mol-ecules will overwhelm the strip. The results will not be representative of the volatiles actually in the debris. If the ignitable liquid concentration in the debris is significantly high, the time and temperature for optimal extraction will be less than what is required from samples with only trace quantities of ignitable liquid. Newman et al. have included an analysis protocol in their paper.[9]

For the Florida Fire Marshal's laboratory, numerous experiments and validation studies have established internal protocols for the use of the carbon membrane. These parameters consistently produce results that approximate the ratios and range of components seen in neat and diluted standards. The carbon membrane is cut to a size of 5 × 20 mm. This narrow strip facilitates placement of the strip into a glass vial where it will be desorbed by an appropriate solvent (carbon disulfide). Safety concerns dictate that the analyst should not deliberately smell the debris. An olfactory evaluation of the evidence should be approached cautiously if at all. However, there are occasions when the analyst will encounter fire debris with noticeable odors of ignitable liquid strongly evident when the container is opened. These are typically exposed to the carbon membrane for 2 to 3 h at 66°C or 16 h at ambient temperature. Samples that are not checked for odor, with slight odors, or without a recognizable odor are exposed to the carbon membrane for 16 h at 66°C.

Currently, bulk quantities of carbon membranes are only available from one commercial source. Activated charcoal may be purchased from a variety of chemical supply firms. In the Florida Fire Marshal's laboratory, similar-sized strips cut from of C18 solid-phase extraction disks have been examined to determine their applicability to the passive headspace technique. The early, unpublished results indicate that they may not be a suitable alternative. They appear to be more efficient at extraction of aliphatic compounds than aromatic compounds. This research is continuing.

For E1413, dynamic headspace, the process may be run in either the positive or negative pressure modes. For both modes, both an inlet tube and outlet tube must be fitted on the evidence container. As the most common container for fire debris evidence is a clean unused metal can, the lid of the can is the usual site for these fittings. The positive pressure mode introduces an inert gas into the heated debris container through one tube, and then allows the gas to exit through the tube containing the adsorbent. The negative pressure mode uses a vacuum line to pull the heated vapors of the debris through an adsorbent tube. An equal volume of air is allowed into the evidence container to keep it from collapsing. An adsorbent tube is prepared by placing a sample of activated carbon or Tenax® between glass wool plugs. The efficiencies of the method are found in the temperature at which the debris is kept during the extraction process, the time that a flow is allowed to pass over the adsorbent, and the velocity of the flow. As was seen with the carbon membranes used in the passive headspace method, the concentration of the ignitable liquid in the debris will affect the optimum time and temperature for the process. Failure to use optimum parameters will cause ignitable liquids to either not sufficiently adsorb or to wash off the adsorbent before it can be tested.

For E2154 SPME, the adsorbent is specifically a sorbent-covered fused silica fiber. The sorbent may be one of several common phases used in capillary columns. These include polyacrylate and polydimethylsiloxane. Mixed sorbents may also be utilized. The fibers can be purchased in various sizes. Key to the ability of the fiber to adsorb components of interest from the debris is the thickness of the sorbent coating on the fiber. The thicker the coating, the better it will adsorb highly volatile components.[11] The SPME technique described in E2154 requires that the adsorbent fiber be in contact with the sample vapors for a much shorter period than other adsorbents in order to achieve similar results. The method is very sensitive and capable of processing samples in a much shorter period than either the passive or dynamic headspace techniques.

There are some issues that must be addressed before this method will see wider usage. The fibers are expensive, but can be reused up to 50 times. This appears to make the fiber more cost efficient. As fully half of these uses would have to be devoted to a blank desorption in order to ensure the cleanliness of the fiber, the cost efficiency achieved would be cut in half. Additionally, there is the issue of how to archive the extract. Simply put, thermal desorption of the strip does not allow for an archived extract. Recent work describes a method whereby the fibers are desorbed with a solvent much in the same way as the adsorbents in either the passive or dynamic headspace methods.[12] Should further studies show this to be a valid technique, the issue of SPME extract archival will be solved.

4.2.7　Desorption

Once the ignitable liquids have been adsorbed and concentrated, it is necessary to remove them from the adsorbent. Once removed from the adsorbent, they can be introduced as a sample into an instrument. This may be accomplished through either thermal or chemical means. Thermal desorption requires that the adsorption media be placed into a vessel where it can be rapidly heated to cause the trapped ignitable liquids to move into the vapor phase. Once in the vapor phase, the sample is swept into the instrument by a carrier gas. Thermal desorption is typically performed on adsorbent tubes from E1413, dynamic headspace, or fibers prepared by E2154, SPME. There are commercial desorption devices that provide consistent and reliable desorption. For E2154, SPME, the injection port of the gas chromatograph may be used to provide thermal desorption and direct introduction of the sample to the instrument.

The other method for desorbing the adsorbent is to wash the adsorbent with a solvent. The solvent strips the adsorbed ignitable liquid components from the adsorbent and carries it into solution. The solution is then introduced into the instrument. Different solvents have differing efficiencies for desorbing an activated charcoal strip (ACS). For carbon membranes and activated charcoal, the solvent thought to have the greatest efficiency for stripping ignitable liquids is carbon disulfide. Carbon disulfide has a notorious reputation for its inherent danger. It is flammable, explosive, toxic, and carcinogenic. To combat these dangers, other solvents have been evaluated,[13,14] including diethyl ether, pentane, and others. None, however, match the efficiency of carbon disulfide in completely removing the trapped ignitable liquids from the carbon strip. It is highly recommended that carbon disulfide, or any solvent, be used exclusively inside an operating fume hood.

In order to reduce the consumption and accidental spillage of solvents while preparing carbon strips for analysis, it is best to equip the solvent bottle in the fume hood with a bottle-top dispenser. This device fits into the bottle sealing against the bottle cap fittings. The unit can then be adjusted to deliver a set volume of solvent. The solvent can be delivered directly onto the vial where the carbon strip has been placed following extraction. The vial can then be immediately capped. This procedure eliminates much of the exposure to solvent vapors. It has the advantage of delivering a set quantity of liquid each time. It also eliminates several sources of potential contamination to the solvent. This is important as the amount of ignitable liquid in contact with the carbon membrane directly affects its concentration in the solvent. For example, two carbon membranes from the same evidence container are placed into two separate vials. One has 300 μl of solvent added and the other has 600 μl. If both strips had the same concentration of adsorbed ignitable liquid, the one with the 600 μl of solvent added will be more dilute.

Considering the above comment, analysts may be encouraged to only add sufficient solvent so that the strip is moistened and there is enough liquid present that an aliquot can be taken up in a syringe. While this may be possible for those who are making manual injections, the volume will be insufficient for most gas chromatographs equipped with autosamplers. Autosamplers move syringes over the vials and plunge the syringe needle into the vial. Because the autosampler is designed to avoid bottoming out in the vial, the needle tip will only reach to a specific depth. For autosampler use, the optimum amount of solvent to add to the vial lies just above this maximum depth. This may produce a less concentrated sample than desired, but is necessary. One method to reduce the volume of the solvent while increasing its depth in the vial is to use a vial with a volume-reducing insert or a conical vial. However, evaporation of the solvent through the septum cap of the vial must be considered.

Because the autosampler and vial selection trays are often located above or near the injection ports, detectors, or ovens of gas chromatographs, one must recognize that the vials will be subjected to mild heating. As the auto sampler may contain 40 to 80 or more slots, there may be a considerable time delay before the last samples are to be injected. This heat plus the time delay could cause the evaporation of solvent so that the volume at the time of injection will be below the needle tip. This will cause an injection that will either be blank or, depending on the tightness of the syringe, a vapor injection. The data collected for the instrument will thus not be reflective of the actual sample.

With Tenax®, care must be given to the temperature of desorption. Tenax® is poly(2,6-diphenyl-p-phenylene oxide) with or without carbon added. It is a porous polymer that has been used for years as an adsorbent for organic species. It is reported that its performance with highly polar compounds is poor, that it does not retain highly volatile compounds, and that it may decompose and melt above 250°C.[15] Even with these factors, Tenax® has found many supporters who report very good results.

4.2.8   Standards

Having a method for extraction of ignitable liquids from debris is meaningless unless the analyst has standard ignitable liquids that can be used for comparison. Building a library or reference collection is critical to fire debris analysis. The construction of this reference library should begin with a careful review of either ASTM E1618 "Standard Test Method for Ignitable Liquid Residues in Extracts from Fire Debris Samples by Gas Chromatography–Mass Spectrometry" or E1387 "Standard Test Method for Ignitable Liquid Residues in Extracts from Fire Debris Samples by Gas Chromatography."[3] The library should include representative samples from each of the ASTM classes referenced in the ASTM test methods. The analyst building the library should be aware that more than one sample of each class is necessary as there may be significant variations within a class. This is especially true of light and medium petroleum distillates, isoparaffinic mixtures, and naphthalinic/paraffinic mixtures. This will require the laboratory to purchase ignitable liquids from a variety of sources including gasoline stations, retail, stores, automotive supply stores, and hardware stores.

In the retail, automotive, and hardware stores, analysts should examine cleaning, polishing, gardening, automotive, and painting products for any item marked as either flammable or combustible. Laboratories may even consider contacting various manufacturers, refiners, and marketers for samples of all the products they sell. Even with a large budget and unlimited storage, it is unlikely that they will be able to collect more than 90% of the potential ignitable liquids sold in their jurisdiction. Additionally, there is a significant potential that not all of the ignitable liquids they encounter were originally available from within their jurisdiction. We live in a mobile society, and when people move they sometimes carry ignitable liquids (as household, hobby, and automotive supplies) with them. The reference collection should not be limited to commercially available ignitable liquids, but should also contain the pure compounds listed as targets in the ASTM standards relating

Table 4.1 Review of Findings from the Florida Fire Marshal's Laboratory

Type	1993–2003 (%)	2002–2003 (%)
Gasoline	30.0	35.0
Petroleum distillates	8.0	6.5
Gasoline/petroleum distillate mix	1.8	2.2
Isoparaffinic mixtures	0.8	0.5

to the analysis and interpretation of fire debris patterns. The library must consist of actual samples of the ignitable liquids and compounds as well as the data derived from analysis by the laboratory's instrumentation.

It is recognized that not all laboratories will have either budget or storage space to acquire and maintain a large collection. To expand on their collection and the knowledge of different ignitable liquids, it is necessary to build a network of fellow fire debris analysts who may be able to assist in identifying the unusual ignitable liquid. An additional tool which is available to the analyst is the book *GC/MS Guide to Ignitable Liquids* by Newman, Gilbert, and Lothridge.[16] Under no circumstance should an identification be made through the use of an outside reference alone. To aid in obtaining additional data and reference standards, the National Center for Forensic Science (NCFS) and the Technical Working Group for Fire and Explosions have developed an ignitable liquids reference collection. Gas chromatographic/mass spectral data can be freely downloaded from the database (www.ncfs.org/databases). Vials containing small samples of these standards can be purchased from NCFS for a nominal fee.

If possible, it is recommended that standards be prepared as both dilutions of neat liquids in an appropriate solvent and as extracts using the extraction method that the laboratory has selected. This will allow the laboratory to identify any difference between the components recovered in their extraction process and the profile of the neat or diluted liquid. Most laboratories run their full battery of standards once a year when a GC column is changed or when a new instrument is installed. While this is sufficient in many cases, it is recommended that fresh standards of the most commonly identified ignitable liquids be analyzed at least once a month. This will avoid any charges that the laboratory is using obsolete standards. The question is, which ignitable liquids are the most common? The review of the findings on the 29,055 samples analyzed by the Florida Fire Marshal's Forensic Laboratory from January 1992 to January 1993 compared to the 3734 samples analyzed from only January 2002 to January 2003 clearly indicate that the ignitable liquids commonly found in fire debris are gasoline and petroleum distillates. The results are summarized in Table 4.1.

One must be aware that in the marketing of solvents and ignitable liquids, economics often drive changes to the formulations that are marketed. A standard of a particular brand of charcoal starter fluid purchased in 1999 may have been a medium petroleum distillate then, but may prove to be an isoparaffinic hydrocarbon mixture if purchased this year. For this reason, the laboratory's reference library requires a periodic update. Do not discard the old reference materials. People often keep containers of ignitable liquids for years. An older standard can often be the differentiation between an ignitable liquid used as an accelerant and an ignitable liquid that was actually inherent to the scene.

One additional standard should be incorporated for use by fire debris analysis laboratories on a regular basis, an internal standard. This is a chemical compound not normally found in ignitable liquids. It will chromatographically elute well after the solvent peak. An aliquot of this material should be added to each sample prior to the sample's extraction. It will provide a marker for several quality assurance measurements. The internal standard can be used as a reference for both retention time and concentrations and account for differences in instrument response on different days. The Florida Fire Marshal's Laboratory utilizes a 3-µl aliquot of 3-phenyltoluene.

4.2.9 Miscellaneous

Many of the above sections focus on the passive or dynamic headspace concentration techniques. The distillation and solvent wash methods described in E1385 and E1386 should not be ignored. While both are destructive techniques that will render the extracted sample unusable for any further testing, there are occasions when the methods may be the optimal choice.

These methods are dependent on the proper use of glassware. For steam distillation, specific glassware can be purchased in kits. This specialty glassware is typically sold with fitted glass joints connecting the various glass components (reaction flask, reflux column, condensing tube, and collection flask). For solvent wash, a variety of beakers and/or separatory funnels are required. The glassware should be scrupulously clean prior to use. Ultrasonic cleaners are quite efficient at getting the glassware sufficiently clean to avoid any carryover of previously extracted solvents. However, ultrasonic cleaner will have the effect of eventually causing the glass to become scored and brittle.

Other items of equipment that will be useful for a variety of reasons include precision pipettes for preparation of solutions and deposition of the internal standard. These may either be fixed or variable volume. Be certain to check the setting of the variable volume before use. We have also found that a precision knife (exacto or surgical) or surgical scissors are useful tools for preparing carbon membranes. For the dynamic method, Pasteur pipettes are an economic alternative to the purchase of specialty extraction tubes. These should only be used if one is going to perform solvent desorption as opposed to thermal desorption. Commercial thermal desorption apparatus will require tubes that fit into the apparatus. For the solvent wash and steam distillation techniques, other tools for rendering debris into smaller pieces so they will fit into the glassware are necessary. These include carpet knives, scissors, chisels, hammers, and saws.

In all instances, tools that are to be reused on multiple samples require careful cleaning to prevent carryover and cross-contamination. With steam distillation, solvent wash and, to some extent, dynamic headspace, this means that a considerable amount of effort must be given to the cleaning process. This may be a contributing factor to the growing popularity and use of ASTM E1412, passive headspace. Besides being simple, sensitive, nondestructive, and relatively noninvasive, the equipment necessary is cheap and disposable. A string and paperclip are all that is needed to prepare a device for suspending the membrane in the vapor space of the sample container. The most expensive item is the carbon membrane or charcoal adsorbent. As this will be used to archive the sample after analysis, it will not require cleaning.

4.2.10 Analysis of the Sample

Once the evidence has been extracted so that any trapped ignitable liquids have been isolated, it is necessary to analyze the sample. Statistics previously cited and found in References 2, 5, and 6 clearly indicate that the choice of instrument has become relatively straightforward. Gas chromatography is almost universally accepted as the choice for separation of the components. For detection of the components, flame ionization detectors have given way to mass detectors. This may not always be the case as new developments in instrumentation are already introducing alternatives.

4.2.11 The Nature of the Analyte

Before proceeding, there needs to be a short review of the nature of the analyte. What is an ignitable liquid and what are the interferences we may encounter? Most ignitable liquids are either petroleum based or petroleum derived. Petroleum-based ignitable liquids have been isolated from crude oil through distillation of straight-run and cracked crude oil. These include the light, medium, and heavy petroleum distillates. Petroleum-based liquids may also be blends where specific chemicals are added to distillation fractions to enhance certain desirable properties. Gasoline fits into this category. However, more stringent environmental requirements may require further processing of gasoline in order to remove certain chemicals.

Petroleum-derived liquids begin as petroleum-based liquids. They are then subjected to various chemical processes at the refinery. One process is derivitization where certain chemical compounds may be converted from one class to another, e.g., normal alkanes being converted to branched alkanes to produce isoparaffininc mixtures. A second option may be filtering a fraction through molecular sieves in order to remove certain classes of compounds. The removal of aromatic compounds from ignitable liquids can be accomplished with molecular sieves. There are other patented and proprietary processes used by refineries in order to produce specialty solvents with properties that satisfy specific markets. Other examples of these types of ignitable liquids are dearomatized distillates and naphthenic/paraffinic mixtures. There are also ignitable liquids that may be produced from sources other than petroleum or isolated through more exacting isolation and derivitization processes. These include turpentine, oxygenated compounds (alcohols and ketones), and aromatic mixtures.

Ignitable liquids may contain normal, branched, or cyclic alkanes. They may contain simple aromatics with alkyl functional groups or complex condensed-ring aromatics. The analytical technique must be able to separate these classes into discrete groupings. The analyst must be able to interpret these patterns and recognize the effects of various factors such as weathering/deterioration. Other interferences will plague the analyst. One is being able to differentiate the components indicative of the ignitable liquid from the additive effects of chemicals produced by the matrix. Matrix effect interference may be categorized as either coming from those compounds inherent to the matrix or from the thermal degradation of the matrix. Many of these compounds are the same compounds that occur in ignitable liquids. They add to and obscure the ignitable liquid compounds of interest. A second, but less common interference comes from microbial degradation.[17,18] *Pseudomonas* bacteria that are present in samples containing soil may feed on certain components of ignitable liquids. If the concentration of pseudomonas is high, relative to the concentration of ignitable liquid in the sample, sufficient components may be eaten to render an analysis inconclusive. The development of instrumentation used in fire debris analysis has been driven, to various extents, by the desire to remove or see past these interferences as well as a desire to see more and more sensitive levels.

4.2.12 Historical Notes

Some of the earliest papers regarding fire debris analysis isolated ignitable liquids by steam distillation and made identification through ignition tests, refractive index, or specific gravity.[19] Spectroscopists attempted to use ultraviolet and infrared techniques[20] to identify ignitable liquids but were confronted with the fact that the instruments at the time did not sufficiently differentiate between ignitable liquid types. Many of the early fire debris extraction and analysis methods were derived from environmental or pharmaceutical analysis methods. The types of detectors that could be used limited liquid chromatography. Only when gas chromatography became available with universal detectors did the identification of ignitable liquids become a viable forensic technique.[21]

4.2.13 Chromatography

Chromatography is simply a separation method whereby complex chemical mixtures are separated based on molecular type, size, or boiling point. The first chromatographic separations involved separation of plant extracts into various components that revealed visible color differences. This was early column chromatography. From this, paper and thin layer chromatography developed. Countercurrent extraction utilized serial liquid–liquid separations to move an analyte from one solvent to another, achieving concentration of the analyte.[22] Liquid chromatography requires passing a liquid mobile phase containing the analyte over a solid stationary phase that retains the analyte for different periods depending on the analyte's attraction to the stationary phase. Again, this is based on the analyte's molecular type, size, or boiling point. Because the separated analyte is contained in a liquid phase, the choice of detector for early liquid chromatographs was limited. These were mostly ultraviolet detectors of fixed wavelength or refractive index detectors.[23]

Gas chromatography vaporizes the analyte and carries it over a solid stationary phase. The separation is dependent on several factors. The flow rate of the carrier gas can be adjusted to give greater separation of the analytes, but will increase the time necessary to completely elute the heaviest component. The temperature of the column can be raised or ramped to improve the speed with which the analyte passes through the column. Most important to the power of chromatography have been developments regarding the diameter and length of the column and the nature of the stationary phase. These changes have resulted in major improvements to separation and resolution of analytes.

Gasoline could only be separated into 15 or 20 poorly resolved peaks using the early (1970s) packed columns, even with optimal carrier gas flow and temperature programming. A 21-ft-long, 1/8-in.-diameter packed column was developed for use at the Ohio Fire Marshal's laboratory, which separated gasoline into 60 to 80 peaks. While this was an improvement, each analysis took approximately 60 min to complete. Longer packed columns were not useful as the pressure drop in the carrier gas flow caused significant tailing of the later eluting components. Eventually, capillary columns began to replace packed columns. The earliest were either wall-coated open tubular (WCOT) columns or support-coated open tubular (SCOT) columns made of glass or stainless steel. The stationary phase no longer filled the tube. Instead, the stationary phase coated the wall of the column (WCOT) or the liquid stationary phase coating was applied to a thin layer of support material bonded to the interior surface of the walls of the column (SCOT).[24] These columns had several performance problems (breakage, separation of coating, etc.) that caused many packed column users to be reticent to change. With the advent of fused silica capillary columns where the stationary phase is bonded directly to the interior of the column, many of the old problems were solved. These have now become the standard of use for fire debris analysis. Depending on the column type, length, carrier flow, and temperature program, it is now routine to separate gasoline into 150 to 250 components. While column technology has made significant improvements, they would have meant little without concurrent improvements in detector technology.

4.2.14 Detectors

Early detectors for gas chromatographs utilized available technology. Thermal conductivity detectors (TCD) measured the change in resistance across a modified wheatstone bridge to "see" the analyte as it passed out of the column.[24] The flame ionization detector (FID) quickly surpassed the TCD as the detector of choice for fire debris analysis and remains a valid choice today. The FID depends on the separated compounds eluting from the end of the column into a hydrogen and air flame. In the flame, the compound is converted to ions and electrons that pass through an electrode gap. As these ions pass the gap, they cause a change in the electrical potential across the gap. This change is measured in millivolts and provides one axis of a graph. The other axis is the time from the moment of injection. The axes together produce a graph of peaks and valleys that represent the components in the ignitable liquid. The graphs produced from the FID have been likened to a two-dimensional view of the components of ignitable liquids. One simply sees that a peak occurs at a specific retention time. The FID does not provide the identity of a particular peak. The utility of the FID is that most ignitable liquids are complex mixtures with components occurring at discrete retention times with specific ratios to each other. Comparison of the complex chromatograms of unknowns to chromatograms of standards prepared under the same conditions allows the analyst to make an identification,[25] provided the concentration of interferences is not greater than the concentration of the compounds of interest. When this occurs, the analyst can use a second gas chromatograph with FID and a column of different polarity to produce a different spread of the components. Another choice became available in the early 1980s and today dominates the field of fire debris analysis. This is the mass detector.

Hewlett Packard (now Agilent) and Finnigan MAT were among the first to market bench-top gas chromatographs equipped with a mass detector. These mass detectors are essentially smaller and scaled-back versions of large research-grade mass spectrometers. Then as now, the choice of a bench-top system centered on the use of either an ion trap or linear quadrupole as the mass analyzer or separator. Mass detectors for gas chromatography are composed of an ionization source followed by a separator/mass analyzer followed by an ion detector. The electron impact source is quite common and uses an electron beam (often from a tungsten filament) to ionize gas-phase molecules. It produces ions by removing an electron from the eluting molecule. Both the linear quadrupole and ion trap require the use of capillary columns as the MS process requires very low pressures in the range of 10^{-5} torr. Vacuum pumps are used to maintain a vacuum across the system. Because of the significantly lower carrier flow effluent from a capillary column, the system is more easily taken to vacuum than it would be if the effluent were from a packed column.

The quadrupole mass analyzer/separator has a configuration of four parallel metal rods arranged so that two opposing rods have an applied positive potential and the other rods have a negative applied potential. As ions produced by the ion source travel the central flight path through the rods, the applied voltages will affect their trajectory. This allows only ions of a specific mass to charge ratio to enter the quadrupole filter. All other ions are excluded because their trajectory is changed. The mass spectrum is produced while the detector monitors the ions coming through the filter, as the voltages on the rods are varied.[26]

The ion trap mass analyzer/separator may be viewed as the three-dimensional analogue of the quadrupole, which is essentially linear or two-dimensional. It consists of a ring-shaped electrode that separates two hemispherical electrodes. This allows ions to be trapped inside a small volume. Altering the electrodes' voltages produces a mass spectrum, allowing ions to be ejected from the trap. The design results in a compact, mechanically simple unit capable of high performance.[27]

A comparison of the performance characteristics of the two designs shows that the ion trap has a considerably higher mass range and can produce unit mass resolution throughout the mass range. The trap can also easily be configured to perform tandem (MS/MS) analyses. The ability of the ion trap to accumulate ions and thus increase the signal-to-noise ratio means that it is more sensitive than the linear quadrupole. One main disadvantage of the ion trap is that it has a limited dynamic range due to space charge effects. With the small size of the area where the ions are accumulated and stored, a large number of ions may lead to a loss of performance. Additionally, because these ions are held in such a small space, intermolecular collisions may result in alterations to the ions held in the trap.[28]

The fire debris extract is analyzed by GC/MS and a total ion chromatogram (TIC) is displayed. Because the instrument has added together all of the discrete fragments within a specific mass range (typically 35 to 350 amu) to produce the TIC, it is possible to deconstruct the chromatogram and obtain displays of chromatograms for either single fragments or summed fragments. The fragments chosen are characteristic of certain classes of organic compounds. For example, the fragment weighing 91 amu is characteristic of toluene and other compounds that produce a C1 benzene fragment. The GC/MS can produce a chromatogram representing only those compounds that contain the 91-amu fragment. This is used as one marker for determination of several ignitable liquids as many of them contain aromatics that produce the C1 benzene. The GC/MS can display summed ions as well. If we desire to see a more comprehensive display of aromatics, we can ask for the compounds that produce fragments of 91, 106, 120, and 134. This chromatogram represents the C1, C2, C3, and C4 benzenes present in the extract. Other fragments can be chosen to determine if the extract contains different classes of compounds. For normal alkanes, 57, 71, and 85 representing butyl, pentyl, and hexyl fragments may be selected. The key to using this tool requires an understanding of organic chemistry and how compounds will fragment in the MS. Once the fragments are represented in a reconstructed ion chromatogram (RIC), they can be individually targeted to show the mass spectrum of the compound.

Both the retention time of the compound and mass spectrum are critical in identifying a compound. The use of a library of standards will allow the analyst to determine the degree to which the unknown matches the standard. In order to determine if a mixture of identified compounds or group of ion profiles indicate an ignitable liquid, the ratios of the compounds and fragments to each other must be examined. Unless there is agreement between the ratios seen in the unknown with the ratios expected in a standard, an identity should not be made. GC/MS has a further diagnostic value with some of the newer ignitable liquids on the market. ASTM, in both E1387 and E1618, established a class of liquids known as dearomatized distillates. The TIC or FID chromatogram of a standard can be easily misidentified as a petroleum distillate. This is incorrect, as the aromatics in the liquid have been reduced by derivitization. Only through comparison of the abundance of the alkanes and the abundance of aromatics can the distinction be seen. Typically, in petroleum distillates the level of the aromatics is only 50 to 100 times less than the level of alkanes. In dearomatized distillates, this difference is that aromatics will be 500 to 10,000 times less.

Other detectors for fire debris analysis could be used, but would likely not improve on the results obtained from either FID or MS. The nature of the analyte is such that detectors that respond to virtually all compounds should be used. Detectors that target specific types of compounds such as the electron capture detector (ECD) may have a utility only when those compounds are present as contaminants in ignitable liquids.

The detector that is provoking discussion among fire debris analysts is the tandem mass detector. As alluded to in the earlier section, most ion traps on the market have the ability to do tandem mass spectrometry. At this time ASTM has not published a comprehensive MS/MS method. The use of GC/MS/MS has been promoted as a method for further discrimination of fire debris extracts than is possible by GC/MS.[29] By selecting characteristic fragments to act as the parent or daughter ion, the sample can be analyzed to reveal a chromatogram that excludes all compounds other than those with the selected ion. This has produced results where the interferences of pyrolysis compounds are eliminated. The technique enhances the sensitivity of detection to lower levels than those seen in GC/MS. The question becomes, "How sensitive do we need to be?" Using GC/MS, and looking at extremely sensitive levels, we now realize that ignitable liquids can be found in many places. Many are inherent to the matrices we test.[4] The technique has a great deal of promise and will require thoughtful consideration in development of a standard for use. Several in the forensic community express the view that the technique is unnecessary and that GC/MS is sufficient. It has been likened to clipping one's fingernails with a chainsaw. It can be done, but if not done correctly the results will be unacceptable. Having been involved with fire debris analysis since 1979, the author recalls the same arguments and analogy when fire debris analysis began to move from GC/FID to GC/MS.

4.2.15   Alternative Methods

At the second annual symposium by the Technical Working Group for Fire and Explosions in 2002, two novel approaches for fire debris analysis were discussed. If we view the introduction and evaluation of GC/MS/MS as the newest evolution for fire debris analysis, either of these approaches may introduce a further evolution. One technique is the use of Fourier Transform Ion Cyclotron Resonance Mass Spectrometry (FT-ICR/MS).[30] Using ultra-high mass resolving power of greater than 1,000,000, with high mass accuracy of less than 1 ppm, and a rapid analysis, the technique has been presented as being able to characterize ignitable liquids to the exclusion of any background interferences. Additionally, the technique is reported to have promise for distinguishing between brands and possibly allowing individualization of ignitable liquids found in one sample (fire scene) to another (suspect's clothes).

The other technique discussed at the meeting, Stable Isotope Ratio Mass Spectroscopy, was presented as the fire debris analyst's equivalent to DNA analysis.[31] The technique identifies the elements in a compound and presents them in terms of the ratios of the various stable isotopes for the elements. Using compound-specific isotope analysis, individual hydrocarbons in two or more fire debris samples could be compared to make a solid connection between the two. Because of the heterogeneity of natural isotope distribution, the probability of two compounds from different sources having the same isotope ratio is unlikely. If multiple compounds are analyzed, the probabilities for individualization are increased. The technique has previously been applied to the geochemical and pharmaceutical industry with validated results.

4.3 Standard Operating Procedures

Standard Operating Procedures (SOPs) are used by most organizations to provide a clear guide for the work they do and how it is to be done. They are used internally to provide clear targets for acceptable performance. Externally, they may be used to allow an audit of the quality of the performance of the organization. While a standard operating procedure can be written to cover almost any topic (I have read "Standard Operating Procedures on how to write Standard Operating Procedures"), they are most useful when written to cover the essential work of the organization. A great deal of care must go into their preparation. In most cases they will not be written as extensions of the personal opinion of the author, but will have a basis in other authoritative materials. The authoritative materials they draw from may include laws, statutes, regulations, accepted industry practices, or standard practices with a wide acceptance. When written, the author must find a balance between specificity and vagary. If the SOP is too specific and detailed, any deviance, regardless of the validity of the deviance, indicates the failure of the organization to follow its SOP. This is a serious charge and can affect the credibility of the organization. Congruently, if the SOP is written to be so vague that it could be interpreted a variety of ways, it fails to provide direction and abnegates its purpose. This also affects the credibility of the organization.

One of my favorite quotes comes from Ralph Waldo Emerson. He said, "What you do speaks so loud that I cannot hear what you say." His meaning is clear. Our actions define who we are. Well-written SOPs give direction. An analogy is that the SOP is like a roadmap. It will show you the routes to take if you want to go from point A to point B. To be useful, the map must be current and accurate. If the author requires that only one method is acceptable, the roadmap will only have one route marked. If there are acceptable alternative routes, they should be included on the map.

For the purposes of this chapter the focus will be on development of SOPs that affect fire debris analysis. Rather than provide templates for SOPs, the references and basis for them will be discussed. It must be recognized that not all laboratories will require identical sets of SOPs. There will be variations in laws in different states that will introduce differences in how evidence is stored and processed. There will be variations in the fire debris extraction and analysis procedures employed by different laboratories.

They are needed to assure the veracity and consistency of forensic analyses. Once written and disseminated, laboratory personnel must be trained to understand what they mean. They should be based on accepted scientific, workplace, and security guidelines.

4.3.1 Historical Development

For many years forensic laboratories developed internal procedures that were seldom codified. Papers were written describing new methods or techniques, but there was no national impetus for creation of consensus standards. Often the analysts with experience in fire debris analysis had a sheaf of notes, chromatograms, and reference materials tucked away in his or her desk that would only be shared internally. In many cases this led to laboratories developing and using their "own" procedures. Often they became comfortable and complacent with these procedures and failed to recognize or adopt improvements in technology or methods. This led to laboratories performing analyses that were circumspect at best or inaccurate at the worst.

As the importance of association began to be recognized, analysts began to compare methods at national meetings and symposia. The Federal Bureau of Investigations offered training in laboratory disciplines to analysts from state and local laboratories around the county. While many took these teachings back to their laboratories and made improvements, there remained a plethora of methods used in the laboratories around the country. Something needed to be done.

In 1977, the Committee on New Development and Research of the American Society of Crime Laboratory Directors received a report on Scientific Assistance in Arson Investigation.[32] In it, the authors assessed the state of the art in arson investigation. They noted the lack of meaningful technology transfer and identified the need for research in scientific techniques. While the use of gas chromatography was noted, the authors noted the lack of positive identification of the components of ignitable liquids by techniques such as mass spectrometry, infrared spectrometry, etc. They further urged that there be more support for research in these areas.

4.3.2 Congress and the National Bureau of Standards

In August 1979 the Federal Emergency Management Agency, U.S. Fire Administration, submitted a report to Congress on the "Federal Role in Arson Prevention and Control." It noted the lack of any arson orientation in most laboratories and that most laboratories used the tools developed for other disciplines. It recommended that the Center for Fire Research of the National

Bureau of Standards (now the National Institute for Science and Technology) develop guidelines to help laboratory personnel to determine if their equipment was appropriate to test arson evidence. Further, the Center for Fire Research should develop these guidelines based on the experience of the Ohio Fire Marshal's Arson Laboratory and other facilities. The goal was to create a manual to be titled "Model Arson Laboratory." The manual would contain guidelines for expanding the role of existing forensic laboratories. The desired impact was an improvement in the quality of arson investigations.[33]

The National Bureau of Standards accepted the task, and in a cooperative effort with the Bureau of Alcohol, Tobacco, and Firearms, the Center for Fire Research began work with an advisory panel of laboratory directors to prepare consensus guidelines for analyzing accelerants. The guidelines were to be sufficiently broad to "allow a chemist considerable freedom to choose an appropriate test method for analyzing the accelerant."[34] The hope was that the guidelines would be developed by standards organizations such as the National Fire Protection Association (NFPA) or ASTM. The purpose would be for laboratory analysts to gain credibility for their work by using recognized procedures. In early 1982, this committee released a preliminary accelerant classification system that it had developed, asking for an evaluation by forensic scientists. The system identified five classes. Class 1 was light petroleum distillates, Class 2 was gasoline, Class 3 was medium petroleum distillates, Class 4 was kerosene, and Class 5 was heavy petroleum distillates.[35] Further, it provided a peak spread for each class, examples of the class, and minimum requirements for class identification. It is interesting to note how this classification system has developed into the accelerant classification system we use today.

4.3.3   The International Association of Arson Investigators

Perhaps one of the most significant items in the development of national standards for fire debris analysis occurred in the March 1988 issue of the IAAI's magazine *Fire and Arson Investigator*. The IAAI Forensic Science and Engineering Committee published their "Guidelines for Laboratories Performing Chemical and Instrumental Analysis of Fire Debris Samples."[36] The guidelines were comprehensive. They began by describing four methods for extraction of fire debris (steam distillation, headspace analysis, solvent wash, and adsorption/elution. They further described the available instrumentation and made recommendations as to its proper use. They then expanded on the previously described classification system. Last, they presented recommendations on how laboratory reports should be constructed and what information they should contain. A review of the article should be required for all analysts as it succinctly presents recommendations that are, for the most part, valid today.

4.3.4 American Society for Testing and Materials

ASTM is an organization with considerable experience in developing consensus standards. Since 1898 this nonprofit organization has aided the U.S. and the world by preparing technical standards that cover a wide variety of issues. They currently offer 75 volumes containing 9957 standards. These standards cover a wide variety of test methods and specifications affecting almost every facet of life including materials (iron and steel, other metals, petroleum products, textiles plastics, etc.) and technology (water and environmental technology, electronics, energy, and medical devices). The standards that affect the fire debris analyst are in Section 14, General Methods and Instrumentation, Volume 14.02 Forensic Sciences, Terminology, Conformity, Assessment, Statistical Methods. The E30 Forensic Sciences Committee developed the ASTM fire debris extraction and analysis methods referenced in earlier portions of this chapter.

ASTM is very careful when it comes to the issuing of standards. They must first be proposed and agreed to by the appropriate committee. The proposal is then balloted and any negative votes must be addressed and resolved. The item will then go back to the committee for approval and must finally be approved by a larger committee. The process is extensive and is designed to garner input and comments from all concerned parties. It is a model of how to develop a consensus standard. Once published, the standard must be reviewed and balloted again in 5 years. This ensures that changes in technology are incorporated into the standards. The E30 Committee has updated and changed many of the standards relating to fire debris analysis over the years. Techniques that have been found to be obsolete are removed.

One of the most important changes recently passed by the E30 Committee was the redesign to the classification systems seen in E1387 and E1618. Because ignitable liquid manufacturers and marketers are notorious for changing formulations without changing product labels and because they may sell the same chemical blend under a variety of product labels, ASTM found it necessary to create a classification system that would allow consistency between laboratories in reporting analytical results. The original classification system contained the following classes:

Class 0: Miscellaneous
    0.1: oxygenated/miscellaneous solvents
    0.2: isoparaffins
    0.3: normal alkanes
    0.4: aromatic solvents
    0.5: naphthenic/paraffinic
Class 1: light petroleum distillates
Class 2: gasoline

Class 3: medium petroleum distillates
Class 4: kerosene
Class 5: heavy petroleum distillates

Over the years it became clear that manufacturers were introducing more ignitable liquids from the miscellaneous class into the marketplace. To address this, the E30 committee changed the system in 2000. The new system no longer depends on a numerical classification, but allows the analyst to first determine a classification based on the predominant components (petroleum distillate, gasoline, isoparaffinic mixture, normal paraffinic mixture, naphthenic/paraffinic mixture, aromatic blend, oxygenate, and miscellaneous), and then, if desired, further differentiate the material into light, medium, or heavy subclasses. Today The ASTM "Test Methods" and "Standard Practices" for fire debris extraction and analysis are the benchmark to which any laboratory standard operating procedure should refer.

4.3.5 Technical Working Group for Fire and Explosions

With funding from the National Institute of Justice and the State of Florida through the University of Central Florida, the National Center for Forensic Science (NCFS) was established in 1997. Part of the impetus for establishment of the center came from a 1996 survey by Dr. William McGee, professor of chemistry at the University of Central Florida. Dr. McGee was also director of the forensic science studies program under the chemistry department. In the survey, Dr. McGee contacted his former forensic science students in order to determine their opinions as to their needs for postgraduation and professional education, training, and standards. Many of the respondents replied that they needed more and better information on the forensic aspects of fire and explosion investigation. At the same time, the National Institute of Justice tasked the NCFS with forming technical working groups to prepare first responder guides for fire and explosion scenes.

The responses from Dr. McGee's original survey as well as the work with the groups developing the first responder guides prompted the NCFS to organize a National Needs Symposium. Fire debris and explosives analysis professionals from across the county gathered in Orlando, FL, in August 1997. During the two days of the symposium, various speakers tasked the group to identify the needs of the fire and explosives analysis communities and prepare suggestions for improvements. The attendees broke into groups to work on specific issues. The recommendations from the fire discussion groups focused on nine problem areas: education and training, evidence recognition and handling, debris analysis methodologies, communication and technology, use of electronic media, research activities, technical working group activities, identifying new technologies, and canine problems. Their recommendations included development of a new classification system for ignitable liquids, identifying field detection systems, universal standards, consistency in terminology and report writing, minimum continuing education guidelines, research into novel accelerants, support of ASTM processes and standards, and identifying new laboratory instruments.

In response to the National Needs Symposium, NCFS created a fire and explosion technical working group. An organizational meeting was held on April 6 and 7, 1998. At this meeting the Technical Working Group for Fire and Explosions (TWGFEX) was established. TWGFEX currently exists with two branches: the scene group, with investigators who respond to fire and explosion scenes; and the laboratory group, with analysts who perform fire debris and explosives analysis. In 2002, the laboratory group was designated as the Scientific Working Group for Fire and Explosions (SWGFEX).

One of the first tasks of TWGFEX was to develop and disseminate a national survey to laboratories performing fire and explosion analyses. A total of 216 responses (35% of survey) were returned. The information came from a broad spectrum of laboratories across the country. The survey confirmed many of the items identified as problems in the National Needs Symposium. When the results were studied, the following issues related to fire debris analysis were identified as tasks for TWGFEX to address: establish a database/reference collection of materials; support for existing ASTM standards and development of new guidelines for fire debris analysis; improvements to proficiency testing; formalization of job descriptions; and identify, improve, and present training. TWGFEX has begun to address those issues and has posted the final versions of its recommendations on its website (www.twgfex.org). The survey results, analyst training outline, standard procedures (which will also be submitted to ASTM), and the ignitable liquids database are available to any laboratory for use in development of its standard operating procedures (SOPs).

4.3.6   The Need for Standard Operating Procedures

The continual improvements to standards regarding fire debris extraction, processing, and training that come from ASTM, TWGFEX, and other organizations provide the nationally accepted and authoritative references that can be incorporated into standard operating procedures. The broadbased national acceptance provides the validity that was so desired by the early practitioners of the science. Accreditation of disciplines within crime laboratories by the American Society of Crime Laboratory Directors Laboratory Accreditation Board (ASCLD/LAB) requires that the SOPs of the laboratory be linked to nationally accepted standards. More and more, the courts are requiring that expert witnesses establish the validity of their opinions and the scientific basis of their techniques. The establishment of SOPs that are based on the standard methods described in previous sections accomplishes this.

4.3.7 Minimum Standards

As a starting point, laboratories performing fire debris analysis need SOPs covering the submission, storage, and disposition of evidence. This will establish the procedures that define the evidence that may be submitted to the laboratory. It will describe the proper container, how it is sealed, and how it is to be documented. The standard operating procedure will describe the proper intake and intralaboratory transfer of the evidence. It will establish the forms to be used by laboratory personnel and the proper methods for completing them. While the primary focus of this standard operating procedure should be on routine evidence, it should also define laboratory procedures for nonroutine evidence. Anyone who has worked in a laboratory knows that there will be those occasions when a piece of evidence will be submitted that does not fully conform to what is normal. Unless there is a defined procedure that will justify acceptance of this piece of evidence, it must be returned unanalyzed to the submitter. So long as the justification is based on sound forensic and scientific principles, the evidence should be reviewed.

The next two SOPs will describe the procedures to be followed in the preparation and analysis of the evidence. These need to explain the extraction procedures to be used, how the samples will be tracked, and how the equipment will be utilized. They will describe the optimum operational parameters of the instrument and the minimum requirements for making an identification. The requirements and procedures for the quality assurance of the samples and instrumentation may either be incorporated into these SOPs or should form a separate procedure. Here, a decision must be made. How in-depth is the SOP to be? SOPs that rewrite an ASTM standard adding precise instructions as to the order of events, wording of notes, and exact amounts of adsorbent or solvent to be used may be too in-depth. An SOP that removes the ability of analysts to make judgments based on their experience and training does not have a realistic view of fire debris analysis. Any deviation from the procedure is considered a failure to follow the procedure. Conversely, an SOP is too vague if it simply refers to a process without referencing its source. Saying, "follow the passive headspace method" or "follow the instructions in the instrument manual" does not provide sufficient direction or include customization of the SOP to the laboratory.

A balance must be reached where the procedures reference accepted protocols and include details as to how that particular laboratory prefers to handle samples. An example of customization can be found in the procedures for naming samples. There needs to be a consistent method for the sequential assignment of a sample reference number to each item of evidence. This is the identifier used to track the evidence. The laboratory's LIMS system may only allow eight characters for samples while the instrument software will allow only six. The laboratory will need to decide how those characters will be used. Will the first two designate the year? Will the last two be the initials of the analyst? The SOP will define the procedure and ensure consistency in its application.

SOPs on the handling of case files and the paperwork to be included in them are essential, as these are the official records of your work on any particular case. Will you designate a set order for the paperwork? Will it be chronological or separated by function? How will you accommodate a request by an outside agent to review your file? All of these issues can be established in the SOP.

Other than the procedures for evidence handling, extraction, instrumental analysis, quality assurance, and records handling, the remaining minimum SOPs would cover safety and security. SOPs can be as inclusive as you wish. You can create them to cover almost any aspect of the laboratory. The most succinct advice is: "If you do it, write it down and if it's written down, do it!"

References

1. Forensic Laboratories: Handbook for Facility Planning, Design, Construction, and Moving, 1998, Office of Law Enforcement Standards, Office of Justice Programs, National Institute of Justice, U.S. Department of Justice, Research Report NCJ 168106.

2. Final Results for Explosive/Fire Debris Analysts (Survey). 1999, Technical Working Group on Fire and Explosives, National Center for Forensic Science, available at http://twgfex.org, Document Library, Lab Survey Results.

3. General Test Methods; Forensic Sciences; Terminology; Conformity Assessment; Statistical Methods, Vol. 14.02, American Society for Testing and Materials International, West Conshohocken, PA, 2002.

4. Lentini J.J., Dolan J.A., and Cherry C., The petroleum laced background, *J. Forensic Sci.*, 45(5): 968, 2000.

5. Summary Report of Test No. 02-536 Flammables Analysis, 2002, Forensic Testing Program, Collaborative Testing Services Inc., available at http://www.collaborativetesting.com.

6. Ignitable Liquid Residue Analysis Test, 2002, Proficiency Testing Program, International Forensic Research Institute, available at http://w3.fiu.edu/ifri/PT.

7. Occupational Exposure to Hazardous Chemicals, *29 CFR 1910.1450, Occupational Safety and Health Administration, U.S. Department of Labor, Washington, D.C.*

8. Standard on Fire Protection for Laboratories using Chemicals, National Fire Protection Association, NFPA 45, 2000, National Fire Protection Association, Quincy, MA.

9. Newman R.T., Dietz W.R., and Lothridge K., The use of activated charcoal strips for fire debris extractions by passive diffusion. Part I: The effects of time, temperature, strip size, and sample concentration, *J. Forensic Sci.*, 41, 361, 1996.

10. Dietz W.R., Improved charcoal packing recovery by passive diffusion, *J. Forensic Sci.*, 35(2), 111, 1990.

11. Chen J. and Pawliszyn J.P., Solid phase microextraction coupled to high performance liquid chromatography, *Anal. Chem.*, 67, 2530, 1995.

12. Harris A.C. and Wheeler J.F., GC/-MS of ignitable liquids using solvent-desorbed SPME for automated analysis, *J. Forensic Sci.*, 48(1), 41, 2003.

13. Lentini J.J. and Armstrong A.T., Comparison of the eluting efficiency of carbon disulfide with diethyl ether: The case for laboratory safety, *J. Forensic Sci.*, 42(2), 307, 1997.

14. Massey D., Du Pasquier E., and Lennard C., Study for Substitution of Carbon Disulfide by Another Solvent for Desorption of Charcoal Strip (DFLEX™) in Analysis of Fire Samples, Paper presented at 13th INTERPOL Forensic Science Symposium, Lyon, France, October 16–19, 2001.

15. *Purge and Trap Overview*, Restek Technical Guide, 1996, available at http://www.restekcorp.com/voa/pandt.

16. Newman R., Gilbert M., and Lothridge L., *GC/MS Guide to Ignitable Liquids*, CRC Press, Boca Raton, FL, 1998.

17. Mann D.C. and Gresham W.R., Microbial degradation of gasoline in soil, *J. Forensic Sci.*, 35(4), 913, 1990.

18. Kirkbride K.P., Yap S.M., Andrews S., Pigou P.E., Kiass G., Dinan A.C., and Peddie F.L., Microbial degradation of petroleum hydrocarbons: Implications for arson residue analysis, *J. Forensic Sci.*, 37(6), 1585, 1992.

19. Brackett J.W., Separation of flammable material of petroleum origin from evidence submitted in cases involving fires and suspected arson, *J. Crim. Law Criminol. Police Sci.*, 46, 554, 1955.

20. Adams D., The extraction and identification of small amounts of accelerants from arson evidence, *J. Crim. Law Criminol. Police Sci.*, 47, 593, 1956.

21. Lucas D., The identification of petroleum products in forensic science by gas chromatography, *J. Forensic Sci.*, 5(2), 236, 1960.

22. Dean J.A., *Chemical Separation Methods*, Van Nostrand Reinhold, New York, 1969, pp. 45–51.

23. Johnson E. and Stevenson R., *Basic Liquid Chromatography*, Varian Associates, Palo Alto, CA, 1978, pp. 270–315.

24. McNair H. and Bonelli E., *Basic Gas Chromatography*, Varian Associates, Palo Alto, CA, 1965, pp. 91–96.

25. Standard Test Method for Ignitable Liquid Residues in Extracts from Fire Debris Samples by Gas Chromatography, E1387-01, Vol. 14.02, American Society for Testing and Materials International, West Conshohocken, PA, 2002.

26. *Finnigan Trace MS Hardware Manual Revision B FM101555*, Technical Publications, Thermoquest, Manchester, U.K., 1999.

27. Allison J. and Stepnowski R., The hows and whys of ion trapping, *Anal. Chem.*, 59(18), 1072A, 1987.

28. Pannell L., Pu Q., Fales H., Mason R., and Stephenson J., Intermolecular processes in the ion trap mass spectrometer, *Anal. Chem.*, 61(22), 2500, 1989.

29. deVos B.J., Froneman M., Rohwer E., and Sutherland D.A., Detection of petrol (gasoline) in fire debris by gas chromatography/mass spectrometry/mass spectrometry (GC/MS/MS), *J. Forensic Sci.*, 47(4), 736, 2002.

30. Rodgers R.P., Blumer E.N., Freitas M.A., and Marshall A.G., Compositional analysis for identification of arson accelerants by electron ionization Fourier transform ion cyclotron resonance high-resolution mass spectrometry, *J. Forensic Sci.*, 46(2), 268, 2001.

31. Jasper J.A., Edwards J.S., Ford L.C., and Correy R.A., Putting the arsonist at the scene: "DNA" for the fire investigator, gas chromatography/isotope ratio mass spectrometry, *Fire Arson Invest.*, 51(2), 30, 2002.

32. Lowry W.T., Stone I.C., and Lamonte J.N., Scientific Assistance in Arson Investigation: A Review of the State of the Art and a Bibliography, Report presented to the Committee on New Development and Research, American Society of Crime Laboratory Directors, June 1977.

33. The Federal Role in Arson Prevention and Control, Report to Congress, Federal Emergency Management Administration, U.S. Fire Administration, Washington, D.C., August 1979.

34. Rudin E., NBS working on guidelines to aid in identification of arson accelerants, U.S. Department of Commerce News, Washington, D.C., May 11, 1981, 29–30.

35. Preliminary presentation of "Accelerant Classification System," AANotes, *Arson Anal. Newsl.*, May 1982, 56–59.

36. Guidelines for Laboratories Performing Chemical and Instrumental Analyses of Fire Debris Samples, Report by the IAAI Forensic Science Committee, Committee on New Development and Research, *Fire Arson Invest.*, March 1988, 45–48.

Analytical Methods for the Detection and Characterization of Ignitable Liquid Residues from Fire Debris

5

JULIA A. DOLAN

Contents

0-8493-7885-0/04/$0.00+$1.50
© 2004 by CRC Press LLC

137

5.1   Introduction

Laboratory capabilities for the detection of ignitable liquids have improved significantly since these first became a matter of interest for fire investigators and forensic science laboratories. Improvements in both extraction technology and instrumental capabilities have resulted in greater sensitivity and an improved ability to differentiate various classes of ignitable liquids and their residues. Historically, several types of analytical instrumentation have been applied to the analysis of fire debris and the classification of ignitable liquids and their residues. Instrumental methods as varied as nuclear magnetic resonance spectroscopy, fluorescence spectroscopy, and second derivative ultraviolet spectrometry have been applied to the identification of ignitable liquid residues; however, the success of these methods never approached that of gas chromatography-based analysis.[1,2,3] Currently, the majority of forensic science laboratories conducting ignitable liquid residue analysis rely on a gas chromatographic (GC) separation.

5.2   Gas Chromatography

The application of gas chromatography to ignitable liquid residue analysis is well established. Numerous publications in relevant literature allude to the widespread use of GC methods in the analysis of fire debris. Voluntary consensus methods published by the American Society of Testing Materials include standards for the application of gas chromatography in conjunction with both a flame ionization detector (GC/FID) and a mass spectrometer (GC/MS) to the analysis of fire debris.[4,5] The primary reason for the nearly universal application of GC methods to ignitable liquid residue analysis is the need for adequate separation of components comprising petroleum-based ignitable liquids. Because the vast majority of ignitable liquids used as accelerants are derived from crude oil, they require separation of the constituent components for proper characterization of the liquid. Gas chromatography with either an FID or an MS is uniquely suited to fill this need.

The separation capabilities of gas chromatography have long been recognized and applied in the field of petroleum analysis. With appropriate column selection and long run times, a product such as gasoline can be separated into nearly 400 resolvable components.[6] It must be recognized, of course, that separation parameters selected must balance the needs of resolution, cost, and time of analysis. For this reason, most working forensic science laboratories will not achieve this level of resolution.

All chromatographic separations are based on migration through a stationary phase via the mobile phase and the fact that chemical compounds having unique properties will migrate at different rates, thereby allowing for separation. In gas chromatography the mobile phase is an inert gas and the stationary phase is found in the column. Several factors will affect a chromatographic system's ability to separate a mixture, most of which are dependent on the composition of the mixture to be separated. Because gas chromatography, unlike many other types of chromatography, utilizes an inert mobile phase, there are two main factors affecting the relative amount of time a component will spend residing in the stationary phase, and therefore its retention time. These factors are (1) the column and (2) the temperature parameters of the system. Knowledge of the types of mixtures to be analyzed is therefore crucial in selecting an appropriate column and temperature conditions.

5.2.1 Column Selection

The primary factors to consider in selecting a column are type of stationary phase, stationary phase thickness, column length, and column diameter. The composition of the phase selected can enhance separations based on differences in chemical properties and is one of the more important factors in developing an appropriate chromatographic system. The best rule of thumb for column selection is that of "like dissolves like." This principle succinctly states that for most common applications, separation of polar compounds is best achieved with polar columns, and separation of nonpolar compounds with nonpolar columns. In fire debris analysis the vast majority of ignitable liquids are hydrocarbon-based, therefore nonpolar stationary phases are recommended. Use of a longer column and a thicker coating of stationary phase will aid in separating difficult-to-resolve components; however, this has the drawback of an increased run time. Due to the inherent complexity of petroleum-derived ignitable liquids, packed columns are not suitable and will not provide adequate separation. The diameter of a capillary column will affect time of analysis, resolution, and capacity. A smaller column diameter will improve resolution and speed of analysis; however a larger internal diameter will increase the capacity of the column. For forensic fire debris analysis, achieving maximum capacity is often less important than improving resolution and time of analysis. For these reasons a smaller diameter is often advantageous in applications of gas chromatography to ignitable liquid residue analysis. In determining the best type of column for a given analysis, each of these parameters must be selected with consideration given to resolution requirements, time of analysis, cost, column availability, and suitability to other applications.

5.2.2 Temperature Conditions

Development of an appropriate chromatographic method is most dependent on the temperature conditions. In general, the constituent components of ignitable liquids are similar to one another chemically and differ primarily in volatility or boiling point. The preferred temperature for a particular separation will allow for components to pass through at a reasonable speed, therefore the optimum temperature should be below the boiling point of the compound, yet high enough to ensure there is an adequate vapor phase. Because petroleum-derived ignitable liquids tend to have a wide variety of boiling points, a single optimum temperature does not exist. To resolve this difficulty, separations of ignitable liquids are most commonly made under conditions of increasing temperature. The use of a temperature program allows for good separation of components having a wide range of boiling points. Selection of injector and detector temperatures is also dependent on the boiling point of the mixture. Careful selection of temperature parameters will result in a separation that meets the resolution requirements without unnecessarily long run times. Development of an appropriate method will depend on the individual needs of the user and the specific application and are often determined by a combination of referencing the work of others and trial and error.

5.2.3   GC Detectors

A variety of detectors may be used in conjunction with a gas chromatographic system. Requirements for the analysis of ignitable liquid residues mandate that such a detector be nearly universal and of adequate sensitivity such that the low levels of ignitable liquids that survive the fire and extraction process can be detected. Thermal conductivity detectors (TCD), while capable of detecting the components of interest, lack the requisite sensitivity. More commonly used are the FID and the MS. Flame ionization detectors offer the advantages of responding to any compound containing organic carbon, making it nearly universal. Additionally, the FID is very sensitive considering its universal nature, reported to have a limit of detection of 10^{-11}g ($\sim$50 ppb).[7] Mass spectrometers, in addition to being both universal and sensitive, have the advantage of providing structural information. A more detailed examination of mass spectrometry and its applications to ignitable liquid residue analysis will follow.

5.2.4   Pattern Recognition

The classification and identification of ignitable liquid residues is based almost completely on the application of pattern recognition techniques to gas chromatographic data. While chromatographic methods are not generally recognized as identification techniques per se, pattern recognition of complex chromatograms is in most cases sufficient to adequately identify an ignitable liquid or its residue. Because petroleum products are derived from naturally

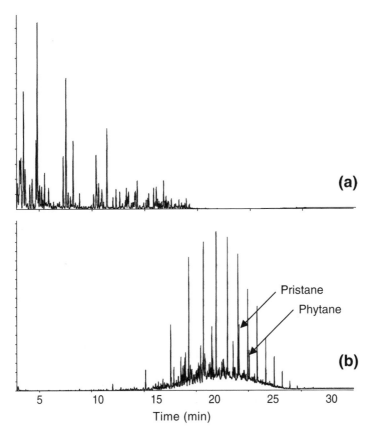

Figure 5.1 Chromatographic patterns of (a) 25% evaporated gasoline and (b) diesel fuel.

occurring crude oil deposits, small groups of compounds present in these products will occur reproducibly in specific ratios leading to recognizable patterns. Figure 5.1 shows chromatograms of two petroleum-based ignitable liquids that are good examples of the utility of pattern recognition. Groupings of peaks in the gasoline pattern (Figure 5.1a) are easily recognized by an experienced fire debris analyst and represent compounds present in gasoline. Each group of peaks is composed of compounds of similar chemical composition and boiling points; therefore, they will tend to elute in a relatively narrow time frame. Each cluster is examined to determine if the relative retention times and peak ratios are consistent with what is expected for gasoline. In addition, peak groups are compared with one another to determine whether the overall pattern is consistent with that of a known gasoline sample. An examination of the diesel fuel pattern (Figure 5.1b) demonstrates that pattern recognition will easily distinguish a heavy petroleum distillate-type pattern from that of a highly refined product such as gasoline. Notable features of the distillate pattern include the overall distribution of peaks, which may be described as a normal or bell-shaped curve, or as a Gaussian distribution. Other important features of this pattern include the C_{19} and C_{20} isomers pristane and phytane, which appear immediately following the normal C_{17} and C_{18} peaks. By obtaining and analyzing a wide variety of reference ignitable liquids, an analyst becomes more adept at recognizing the critical features of a pattern that may be indicative of a petroleum product. Pattern recognition is still appropriately applied even in the presence of additional chromatographic peaks present due to products of the thermal decomposition and partial combustion of various items of fire debris. More detail on the classification and identification of ignitable liquids by gas chromatography is provided in the voluntary consensus standard document "ASTM E 1387-01 Standard Test Method for Ignitable Liquid Residues in Extracts from Fire Debris Samples by Gas Chromatography."[4] Even when a detector capable of providing structural information such as a mass spectrometer is used, recognition and comparison of patterns is still necessary for identification of most common ignitable liquids.

5.3 Gas Chromatography/Mass Spectrometry

The advantages offered by gas chromatography can be enhanced by coupling the separation system with a detector designed for identification of chemical compounds. By combining the separation capability of a gas chromatograph with the ability to provide structural information offered by a mass spectrometer, the technique of gas chromatography/mass spectrometry (GC/MS) offers a significant improvement in the analysis of ignitable liquid residues and has become the preferred technique for this application. There are two major ways in which the fire debris chemist can take advantage of the additional information available from a GC/MS system. One obvious advantage is that the identity of nearly any compound present in the unknown mixture can be determined. This may be important when single component ignitable liquids are present or in cases where an ignitable liquid is comprised of so few components as to make pattern recognition unsuitable. The other key benefit to using GC/MS in the analysis of ignitable liquids is the ability to use structural features that represent various compounds of interest in conjunction with retention time data and pattern recognition. Techniques such as target compound chromatography and extracted ion profiling rely on both structural information acquired by the mass spectrometer and the separation of components achieved by the gas chromatograph. Both of these methods of analyzing data provide a great deal of additional information to the fire debris analyst and are very useful tools. Their application to the analysis of

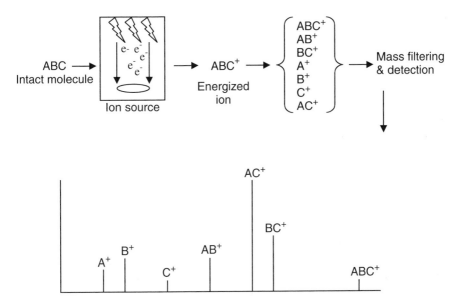

Figure 5.2 Schematic representation of what occurs in a mass spectrometer. Intact ions enter the source, where they are ionized. Excited ions may then fragment. Molecular ions and fragment ions are next subjected to mass filtering and detection, resulting in the graphical representation referred to as a mass spectrum.

ignitable liquid residues will be discussed following a brief overview of the fundamentals of mass spectrometry.

5.3.1 Fundamental Theory of Mass Spectrometry

Mass spectrometry is an analytical technique capable of providing a wealth of structural information. This structural information is derived from the organized data resulting from the systematic ionization and fragmentation of compounds entering the mass spectrometer system. As a chemical species enters the mass spectrometer, several things occur. These discussions will focus on electron impact (EI) mass spectrometry, as that is what is most useful for fire debris applications. A schematic representation of how a mass spectrometer functions is depicted in Figure 5.2. The first step is the ionization and fragmentation of the molecules. As a molecule enters the ion source it is bombarded with electrons, which results in the formation of an energized ion. This energized ion may fragment due to its excess energy or may remain intact as a molecular ion. Following ionization and fragmentation is mass filtering. Mass filtering is the process by which the ions are sorted, based on their mass-to-charge ratio (m/z). Once the ions have been sorted, they are then detected and the data is tabulated and organized into a chart showing abundance vs. mass — the mass spectrum. The information contained in a mass spectrum provides a great deal of structural information and is sufficient for identification in most cases.

The key to mass spectrometry is that the fragmentation of the species will be consistently reproducible when the mass spectrometer conditions are the same. Many mass spectrometer settings will affect how fragmentation will occur, and have therefore been standardized. For example, all common mass spectral libraries are comprised of spectra obtained from ion sources operating at 70 eV. Because high source energy would result in much greater fragmentation than a lower energy, it is important that these conditions be standardized in order to have spectra valid for comparison. Fragmentation is also affected by the structure of the original molecule, its stability, and the stability of the fragments produced. For example a molecule such as naphthalene, a fused-ring aromatic compound, would be expected to be very stable and to show little fragmentation. This is confirmed by examining its spectrum (see Figure 5.3a). The mass spectrum of naphthalene has as its base peak, or most abundant ion, the molecular ion occurring at m/z = 128. Other molecules such as alkanes are much more prone to fragmentation and will have a greater total number of fragments and only a small abundance of the molecular ion (see Figure 5.3b). These fundamental properties of the compound being analyzed will not change; consequently, fragmentation will be consistent.

Mass filtering, or the process of sorting out ions based on their mass-to-charge ratio, can be accomplished by a variety of means. Each type of mass spectrometer has advantages and disadvantages. Quadrupole and ion trap instruments are most commonly used in the field of fire debris analysis. The quadrupole offers the advantage of being an economical and robust instrument that is both compact and user friendly. The ion trap can offer great sensitivity and high resolution in a compact, relatively inexpensive instrument. For fire debris applications high resolution is not necessary, nor is a particularly wide mass range. For these reasons, fire debris analysts can take advantage of these easy-to-operate, low maintenance, and economical instruments. The final steps in obtaining a mass spectrum are detecting the ions, and organizing and presenting the data in a usable form.

Mass spectrometers may be operated in one of two primary modes: full scanning or selected ion monitoring (SIM). Each method has its advantages and disadvantages and is more appropriate for specific applications. When operating in the full scanning mode, the mass filter of the mass spectrometer is set such that it collects all m/z values within a given range. That is, it scans from a high value to a low value and collects data at each distinct m/z value between the upper and lower limits. The parameters for a selected ion monitoring method are designed such that data is only gathered for a few particular ions. Data for ions other than the designated selected ions is never acquired

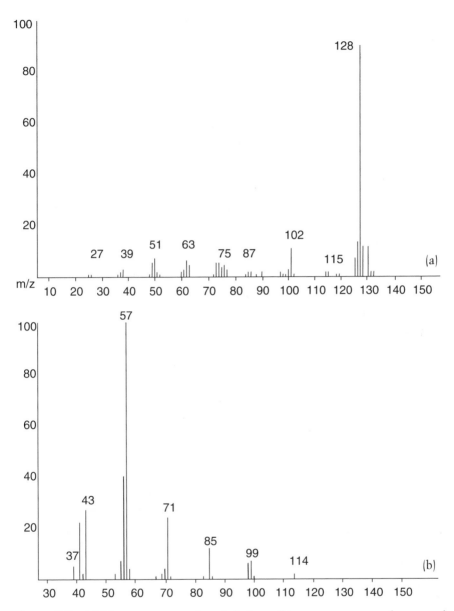

Figure 5.3 (a) Mass spectrum of naphthalene; (b) mass spectrum of a normal alkane (n-octane).

by the data system. Although fragments may be created at m/z values other than those selected, data regarding those ions is never recorded. Scanning is a more general technique and is good for screening and general unknowns. Because it detects all ions within the specified range, it is unlikely that a compound will not be detected. In addition, because a full range of m/z data is collected, one can examine a full mass spectrum for any point in time of the chromatographic run. In this manner one can obtain structural information for any component of the mixture. The downside to full scanning is that it is a fairly inefficient technique and lacks sensitivity. Consider, for example, a scanning run set with a range of 300 amu. An equal amount of time is spent looking for each ion within the specified range. This means that because a single scan must collect data for 300 different m/z values, it is spending less than 1/300 of its time on any one m/z value. These conditions clearly do not favor sensitivity. Selected ion monitoring, however, allows the user to specify one or several specific ions for which the instrument parameters will be set. In this manner, less time is wasted looking for ions that are not of interest. For example, if three ions are selected, then the method will spend nearly 100 times more time looking for each ion than a scanning method with a range of 300 amu. In order to specify appropriate ions, however, one must have good knowledge of the sample. In cases in which one is dealing with a true unknown, SIM would not be appropriate due to the fact that one would not be able to specify appropriate ions. SIM is often used for quantitation of specific compounds when other components of a sample are not of interest. It is a much more sensitive technique, especially when a single ion is used. The technique has no value; however, when dealing with true unknowns or when one desires full mass spectra. Fire debris analysts should always use a full scanning method for the initial analysis. Based on information obtained from the scan data, it may be useful to gather additional data using SIM methods.

5.3.2 Fragmentation of Classes of Compounds Significant to Fire Debris Analysis

To understand how fire debris analysts can take advantage of the structural information provided by the mass spectrometer and use it in conjunction with traditional chromatographic pattern recognition, one must first have a basic understanding of the types of ions that are representative of the various types of compounds encountered in fire debris analysis. A good understanding of the composition of crude oil and the products derived from it will aid in this endeavor. Additionally, a review of basic organic chemistry and the structural elements and/or functional groups that define the various classes of compounds will be helpful in predicting the types of ions expected.

Petroleum-derived ignitable liquids are generally comprised of compounds that can be classified into one of five general categories: alkanes (including normal and branched-chain isomers), cycloalkanes, and aromatics, which are further sub-classified into simple aromatics, indanes, and polynuclear aromatics. Other types of compounds may be of interest due to the fact that they often appear in debris samples as either products released

Table 5.1 Classes of Compounds and Common Fragment Ions

Class of Compound	Empirical Formula	Example Structure	Typical Ions
Alkanes (paraffins)	C_nH_{2n+2}		43, 57, 71, 85, 99 (+14)
Alkenes (olefins)	C_nH_{2n}		41, 55, 69, 83 (+14)
Cycloalkanes (cycloparaffins, naphthenes)	C_nH_{2n}		41, 55, 69, **83** (+14)
Alkylbenzenes (simple aromatics, benzene-based aromatics)	$C_{6+n}H_{6+2n}$ [where n corresponds to the alkyl chain(s)]		91, 105, 119, 133
Polynuclear aromatics (PNAs, PAHs, naphthalenes)	$C_{10+n}H_{8+2n}$ [where n corresponds to the alkyl chain(s)]		128, 142, 156
Indanes	$C_{9+n}H_{10+2n}$ [where n corresponds to the alkyl chain(s)]		117, 118, 131, 132, 145, 146
Styrenes	$C_{8+n}H_{8+2n}$ [where n corresponds to the alkyl chain(s)]		104
Terpenes	$(C_5H_8)_n$ [where n ≥ 2] (most commonly encountered are monoterpenes, n = 2)		93, 136

from the substrate during the extraction process or as products of pyrolysis or partial combustion. The most commonly encountered of these compounds include alkenes, styrenes, and terpenes. Nonhydrocarbon compounds may be encountered in fire debris analysis but with significantly less frequency than the types of hydrocarbons described above. Most often these would include oxygenated compounds such as alcohols, aldehydes, ketones, and esters. Table 5.1 summarizes some of the important ions associated with the various hydrocarbons of interest.

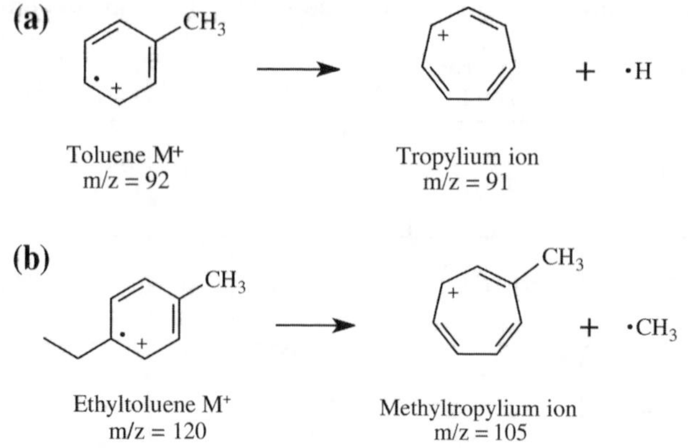

(a)

Toluene M+
m/z = 92

Tropylium ion
m/z = 91

+ •H

(b)

Ethyltoluene M+
m/z = 120

Methyltropylium ion
m/z = 105

+ •CH₃

Figure 5.4 Tropylium ions are predominantly formed when there is a single substitution on the benzene ring, whereas substituted tropyliums are favored when there is more than one alkyl substitution: (a) Formation of the tropylium ion (m/z 91) from toluene and (b) formation of a methyltropylium ion (m/z = 105) from ethyltoluene.

For many of the compounds of interest, determination of the fragment ions is intuitive. Alkanes will produce alkyl chains, as will alkenes. In addition, alkenes will produce similar fragments that include a double bond, thereby appearing with m/z values that are two less than their alkyl counterparts. Polynuclear aromatics, terpenes, and styrene-based compounds will be best represented by their respective molecular ions due to their inherent stability. Indane compounds are also fairly stable due to their fused ring structures; therefore, their respective M-1 ions and molecular ions are well represented in their spectra. The alkylcyclohexanes are the most commonly seen cycloalkanes and are indicated by the 83 ion representing the cyclohexyl group. This is because the cleavage of two bonds is necessary to produce a fragment ion from a cyclic alkane. It is a worthwhile exercise for the reader to examine several spectra representing the various classes and to elucidate the structures of the principal fragments in order to better understand the process.

The one commonly encountered type of compound that exhibits an abundant fragment ion with a less obvious source may be the alkylbenzenes or simple aromatics. As expected, alkylbenzene compounds will show a significant molecular ion; however, their spectra are most often dominated by a peak at m/z 91. This peak is due to the formation of the tropylium ion. The tropylium ion is most prominent when there is but one alkyl substitution on the benzene ring; substituted tropylium ions are more abundant when there are additional alkyl groups. Figure 5.4 shows the formation of the tropylium ion and its methyl-substituted analogue from the toluene and ethyltoluene molecular ions. Tropylium and substituted tropylium ions are formed from an alkylbenzene when the alkyl chain is cleaved to form a radical and the carbon adjacent to the ring collapses into the ring to form the tropylium (or substituted tropylium) ion. These ions are remarkably stable and by remembering that one of the main factors affecting fragmentation is the stability of the resulting ion fragments, one can understand the great abundance of these fragments in mass spectra of alkylbenzenes. Although it may not be obvious, tropylium ions exhibit their particular stability due to aromaticity. Although a seven-membered ring, the tropylium ion carries a positive charge, and therefore has 6 pi electrons, thereby meeting the requirements of the Debye–Hückel rule for aromaticity. There are also numerous resonance structures further contributing to the stability of this ion. Generally speaking, the longest chain on a multiply substituted alkylbenzene will be cleaved because this cleavage results in the formation of larger and therefore more stable free radicals. Understanding the formation of the tropylium and substituted tropylium ions allows us to understand why a peak at m/z 91 will be prominent in toluene, ethylbenzene, and propylbenzene, and though it will be present in the xylenes and ethyltoluenes, the abundance of the m/z 105 ion will be greater. Understanding the reasons how and why specific ions represent certain classes of compounds can significantly enhance a fire debris analyst's ability to interpret complex data. While a more detailed discussion of fragmentation is beyond the scope of this book, it is recommended that the reader refer to a text on mass spectrometry or interpretation of mass spectra for a more in-depth discussion of this topic.

5.4 Approaches to GC/MS Data Analysis

There are several ways in which mass spectral data can be used to assist the fire debris analyst. In cases in which there are only a few components of an ignitable liquid, the analyst may examine the full spectra of the peaks of interest and use this information to identify the individual components of the mixture. This is often necessary when ignitable liquids such as single component ignitable liquids, oxygenated products, or normal alkane products are present. Identification of common pyrolysis products may also assist the examiner when no recognizable ignitable liquid pattern is present, yet there is a complex chromatogram. More often, however, the analyst does not use the mass spectrometer to absolutely identify compounds, but rather uses spectral features characteristic of various classes of compounds in conjunction with basic pattern recognition techniques to analyze data. The three common ways in which analysts utilize spectral features are (1) target compound chromatography, (2) extracted ion profiling, and (3) library reports. The theory and application of each of these techniques will be discussed, with an emphasis on extracted ion profiling.

5.4.1   Target Compound Chromatography

Target compound chromatography is a data analysis technique designed to aid in the identification of ignitable liquids when they are present in relatively low abundance with respect to coextracted compounds derived from the sample matrix. More simply put, it is designed to find a needle in a haystack. This technique involves searching data for target compounds based on retention time and spectral characteristics. The target compounds are specific chemical components previously determined by the user and input into the software. In searching the data for the presence of these compounds, the software examines a peak for retention time and determines if it has the previously specified ions of interest. Generally speaking, to meet the criteria to be considered a target compound, the unknown compound must fall within the specified retention time window, have the correct target ion, and have the specified qualifier ions with normalized abundances falling within the specified range. The target ion is often the base peak but may be another significant peak. Qualifier ions are other peaks present in the reference spectrum of the target compound that are used to minimize the possibility of erroneously characterizing incorrect compounds as target compounds. The criteria for identifying a target compound — retention time, target ion, and qualifier ions — are variables specified by the user based on data of reference target compounds. This information is then input into the data analysis software as part of the target compound method. The selection of appropriate criteria for target compounds is crucial due to the fact that many hydrocarbon isomers will have similar spectra.

The selection of compounds to be used as target compounds is critical to the effective use of target compound chromatography. When considering that many of the ignitable liquids of interest are composed of hundreds of individual components, the process of selecting a relatively few compounds that are capable of representing that product is a formidable task. Useful target compounds generally must meet two important criteria: they must be consistently present in the ignitable liquids of interest, and they should not be commonly encountered from sources other than from refined petroleum products. Research conducted by Keto and Wineman established suggested lists of target compounds for various classes of ignitable liquids, including gasoline, medium petroleum distillates, and heavy petroleum distillates.[8] These target compounds have been adopted into the general consensus standard method for fire debris analysis by gas chromatography-mass spectrometry.[9]

Target compound chromatography, like other methods of ignitable liquid residue data analysis, still chiefly relies on pattern recognition techniques. The TCC computer program isolates compounds meeting the criteria of the predetermined target compounds and a semiquantitation of these target compounds is performed based on the target ion. From this data — retention time and semiquantitative abundance — a target compound chromatogram is constructed. A target compound chromatogram is not a true chromatogram but rather a visual representation of the data included in a chromatogram — retention time and abundance — in the form of a stick plot. Most GC/MS software will perform target compound methods but do not include the ability to create this pseudo-chromatogram. Therefore, in order to be able to utilize pattern recognition techniques, the target compound information is transferred into a spreadsheet program and the plot is created from there. The target compound chromatogram is then compared to one of a reference ignitable liquid created in the same manner. As in all methods of data analysis for ignitable liquid classification and identification, the unknown pattern is visually examined and compared with data obtained from a known sample. This technique offers the ability to find components of interest in a highly complex chromatogram in which contributions from pyrolysis are considerable. The main disadvantage to using target compound chromatography lies in the fact that the reconstructed stick plots do not have the same appearance as a traditional chromatogram, so reading the data requires additional practice and training in the method, even for an experienced analyst.

5.4.2   Extracted Ion Profiling

Extracted ion profiling, also referred to as reconstructed ion chromatography or mass chromatography, is the most commonly used method of using mass spectral data characteristics in conjunction with chromatography to analyze ignitable liquid data. Much like target compound chromatography, it allows the analyst to collect full spectral data throughout a chromatographic run and to focus on compounds of interest by electronically "filtering out" other interfering components from the chromatogram. Unlike target compound chromatography, however, extracted ion profiling does not focus on a limited number of preselected compounds. Rather, it focuses on mass spectral characteristics that are common to particular classes of compounds. This data is then used to create extracted ion chromatograms or extracted ion profiles indicative of certain types of compounds. These EICs or EIPs can then be compared to those obtained from reference ignitable liquids.

It is important to understand the differences between selected ion monitoring (SIM) and extracted ion profiling. Recall that selected ion monitoring is a technique in which data is collected for only a few predetermined ions of interest. This technique is much more sensitive for compounds of interest than is scanning; however, it does not collect full spectral data. Consequently compounds present in the mixture that do not have a significant abundance of the monitored ions will not be detected. In addition, because SIM only collects data for a relatively small number of ions, mass spectra cannot be searched against libraries or used as a basis of identification. In contrast, the technique of extracted ion profiling collects data in the full scanning mode; therefore, data is acquired for a broad range of ions. This ensures that unexpected components will be detected. In addition, a full mass spectrum is available that is suitable for comparisons with both libraries and other reference spectra. It is from this full scan data that data representing significant classes of compounds is extracted or isolated. By doing this, one can focus on a particular ion or group of ions without losing data related to other ions. The technique of extracted ion profiling is then able to offer the advantages of not omitting any important data while still providing a mechanism to minimize contributions from interfering compounds, allowing the analyst to focus on data of potential significance.

While the terms are often used interchangeably, in this chapter the term "extracted ion chromatography" refers to focusing on a single ion of interest, whereas "extracted ion profiling" involves focusing on several related ions of interest. To create an extracted ion chromatogram, the software displays the abundance of a specific ion in the form of a chromatogram. Rather than a total ion chromatogram, which is representative of the abundance of all ions detected, the extracted ion chromatogram shows only the abundance of the specified ion vs. time. This technique effectively filters out components that do not contain the ion of interest, allowing the analyst to examine a much simpler chromatogram. The process of summing several extracted ion chromatograms results in an extracted ion profile. The key to extracted ion profiling lies in the fact that related compounds will have spectra with similar features. Because of this, ions that are abundant in a specific class of compounds can be used to represent that class of compounds. Subsequent to selecting several ions to represent a class of compounds, one sums the abundance of each of the ions, resulting in a single chromatogram referred to as an extracted ion profile. As with the total ion chromatogram, these extracted ion chromatograms and profiles are suitable for comparison with similarly obtained data from reference ignitable liquids.

To illustrate the techniques of extracted ion chromatography and extracted ion profiling, consider an example. In the discussion on fragmentation, it was seen that fused ring aromatic compounds are relatively stable and therefore have strong molecular ions in their spectra. Knowing this, if we want to look at a complex chromatogram such as gasoline and focus on the naphthalene and substituted naphthalene compounds, we would make select ions such as 128 (mol wt of naphthalene), 142 (mol wt of the methylnaphthalenes), and 156 (mol wt of the C_2-substituted naphthalenes). Figure 5.5a–c show the extracted ion chromatograms for ions 128, 142, and 156, respectively. Compare the complexity of these extracted ion chromatograms with the total ion chromatogram of gasoline shown in Figure 5.5e. By examining the appropriate EIC, one can easily locate the desired naphthalene compound.

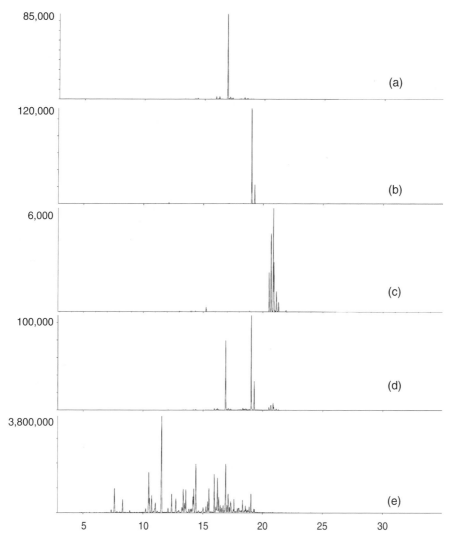

Figure 5.5 Data from an evaporated gasoline sample. (a) m/z = 128 Extracted ion chromatogram; (b) m/z = 142 Extracted ion chromatogram; (c) m/z = 156 Extracted ion chromatogram; (d) Extracted ion profile representing the polynuclear aromatic compounds ("PNA Profile") consisting of the sum of the 128, 142, and 156 ions; (e) Total ion chromatogram.

The extracted ion profile utilizes the same concept; however, rather than creating individual chromatograms for each ion, the profile summarizes the data contained in multiple individual extracted ion chromatograms. By grouping together ions that represent a single class of compounds, one can sum the extracted abundance data and use a single chromatogram to represent a particular class of compound. Figure 5.5d shows the polynuclear aromatic compound extracted ion profile that is created by summing the individual 128, 142, and 156 EICs. In a manner similar to that used with single ion chromatograms, the analyst is able to focus on a single class of compounds while effectively filtering out other components of the mixture through the use of a summed profile.

The differences between these two techniques are fairly minor, and use of one over another is generally a matter of the preference of the individual analyst. Fewer chromatograms result when the summed profiling technique is used — generally in the range of 4 to 5, rather than the 10 to 15 usually generated when single extracted ion chromatograms are used. In addition, when profiling is used, all peaks will share a common abundance scale. This has both advantages and disadvantages. The advantage is that by having a common scale all the components will appear in proportion to one another, making it easier for the analyst to conduct a visual pattern recognition examination and compare ratios of the various components present. The disadvantage is that minor component patterns are not as easily seen, as they may often be nearly lost in the baseline when a common scale is used. Compare the pattern of the C_2-naphthalenes shown in single extracted ion chromatogram of Figure 5.5c with their pattern in the summed profile shown in Figure 5.5d. As relatively minor components, the C_2-naphthalenes are more easily seen in a single EIC, where the scale is adjusted to their abundance. Another factor to consider is that when summed profiles are used, there is often an increase in signal-to-noise ratio. Examine the individual extracted ion chromatograms and the summed profile for the simple aromatic compounds shown in Figure 5.6 to see how the summing of the individual ions results in slightly greater sensitivity. The summed profile shown in Figure 5.6e also displays data having ratios more consistent with TIC and FID data, making it a more recognizable pattern. The fact that a summed profile better represents the overall pattern is a significant benefit to the use of summed profiles. Whether single EICs or summed EIPs are used, it is imperative that comparisons be made to data acquired and presented under identical conditions. Because there are advantages to both methods many analysts use a combination of both single and summed techniques.

When using extracted ion methods for the analysis of fire debris there are several general guidelines to which the analyst should adhere. Of primary importance is selection of the ions to be used. To develop an appropriate list of ions to be used one should have a general understanding of the composition of refined petroleum products and the expected mass spectra. This knowledge can be used along with published references that list ions significant in fire debris analysis to develop a list of appropriate ions. The entire concept of extracted ion methods is dependent on the use of appropriate ions. Should this method be applied with unsuitable ions, the discriminating

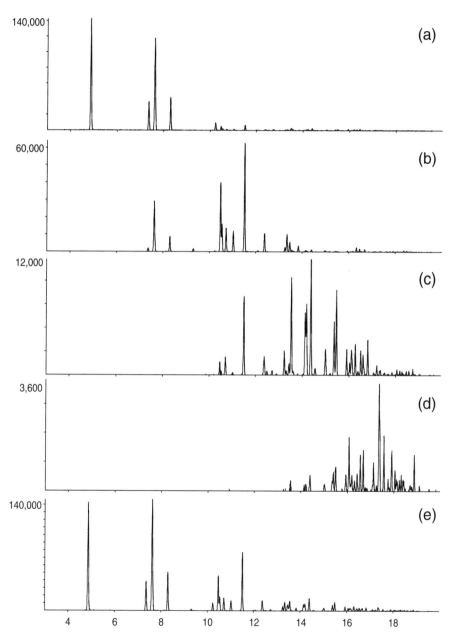

Figure 5.6 Data from a 25% evaporated gasoline sample. (a) m/z = 91 Extracted ion chromatogram; (b) m/z = 105 Extracted ion chromatogram; (c) m/z = 119 Extracted ion chromatogram; (d) m/z = 133 Extracted ion chromatogram; (e) Extracted ion profile representing the simple aromatic compounds ("Aromatic Profile") consisting of the sum of the 91, 105, 119, and 133 ions.

value of the technique will be lost. Another important guideline to remember is that when using extracted ion chromatograms/profiles to identify an unknown, the chromatograms of the unknown must be compared with analogous chromatograms obtained from a known reference source under identical conditions. The comparison process will include general pattern recognition and observation of peak ratios, and must also examine the overall abundances of the various chromatograms so that relative intensities can be compared. Finally, it must be remembered that even with all the emphasis placed on extracted ions, the total ion chromatogram is still the most important data available to the fire debris analyst.

5.4.3 Library Searches

Electronic databases of reference spectra can save a significant amount of time and offer considerable advantages in the analysis of complex data. While one must be extremely wary of using a library as a means of identification, a well-stocked library can provide a great deal of information regarding the components of an unknown mixture. When a major single component is present in a chromatogram, a search of an appropriate computer library can often provide the identity of the unknown compound. Although the quality of spectral libraries has improved greatly since their inception, errors do still exist within them. For this reason, one should never rely solely on a library match to establish the identity of a compound. The analyst may use his or her knowledge of fragmentation in conjunction with a high-quality match to feel secure with the identity of a component, but it is always recommended that if one intends to identify a component, it should be done via comparison to a full spectrum standard from a reliable source. In-house standards run under identical conditions are the ideal, allowing for comparison of both retention time and spectra.

Libraries are also very effective for getting a general idea regarding the class of a particular compound. Because the spectra of different classes of hydrocarbons tend to have common features, the spectra of chemically similar components will have similar spectra. This can sometimes hinder the process of identification, particularly in the case of isomers; however, establishing a class of compound or major structural features can usually be accomplished. Because petroleum-based ignitable liquids contain numerous resolvable components it is often not feasible, nor is it necessary, to identify each individual component. In cases of complex chromatograms it may be useful to run a program that performs a quick library search on all integrated peaks. Most GC/MS data analysis software packages are equipped to do this with minimal user intervention. By getting a brief listing of the possible identities of compounds, one can get a feel for the overall composition of the mixture. For example, the library search report of an isoparaffinic product will list compounds that are exclusively isoparaffinic in nature. While the report would not be providing the absolute identities of the components, it easily indicates that they are branched-chain alkanes. Similarly, when one examines the report for a debris sample, often one will see various common pyrolyzates. Although the use of a library in this manner must be accompanied by more traditional pattern recognition techniques, it can reinforce the conclusions reached.

5.4.4   Summary

There are many ways to examine data. Each has its advantages and disadvantages. Selection of one technique over another is often a matter of examiner preference. An examiner must always utilize the total ion chromatogram and may utilize the additional information provided by target compound chromatograms, extracted ion methods, and library search reports. Often, more than one approach is used to analyze complex data.

5.5   Criteria for Identification of Ignitable Liquids and Their Residues

The identification or classification of an ignitable liquid by GC/MS must meet certain criteria in order for the identification to be deemed scientifically reliable. The process of refining petroleum and subsequent processing of petrochemicals results in numerous classes of liquids that can be grouped together based on similarities in chemical composition. The ASTM consensus methods recognize eight general categories of ignitable liquids, each of which is defined by specific criteria related to chemical composition.[9] The eight major classes defined in the peer developed ASTM methods include gasoline and seven categories that can be further described by their relative volatility as light, medium, or heavy. In addition, a miscellaneous class exists due to recognition of the fact that no classification scheme can be all-inclusive. This section will focus solely on the data requirements for identification as one of these classes; the classification scheme itself is described in greater detail in the next chapter.

In order to identify or classify any ignitable liquid, it is necessary to have an appropriate reference for comparison. Chromatographic pattern recognition is a key factor in all ignitable liquid identifications; therefore, a suitable reference collection representing ignitable liquids having a variety of boiling point ranges and chemical compositions is essential to this endeavor. In addition, a suitable reference collection should contain liquids in various stages of evaporation in order to mimic the effects of fire. Data should be obtained for each of these liquids under appropriate conditions (refer to this chapter's sections on column selection and temperature programming) for comparison with data obtained from unknown liquids or extracted residues obtained under comparable conditions.

5.5.1   Gasoline

Gasoline is derived from crude oil and undergoes significant refining operations, resulting in a product that is rich in aromatics and ranges from approximately C_4 to C_{12} in its unevaporated state. The gasoline category includes all brands of gasoline, including gasohol.[5] A gasoline pattern will have a specific reproducible pattern of aromatic compounds that is easily recognized in the total ion chromatogram. A fresh gasoline pattern will also exhibit substantial isoalkanes in the early portions of the chromatogram along with other aliphatic compounds, although the benzene-based aromatic compounds will dominate the pattern beyond the C_7 portion of the chromatogram. As gasoline evaporates, the overall pattern will appear to shift toward the right, as the more volatile compounds are preferentially consumed (see Figure 5.7). To identify gasoline the overall pattern present in the total ion chromatogram must be consistent with the patterns obtained from known reference samples. ASTM 1618 specifically requires that the C_3-alkylbenzenes *m*-, *p*-, and *o*-ethyltoluene, and 1,2,4-trimethylbenzene must be present.[5] In addition, extracted ion profiles representing the various classes of compounds commonly encountered in petroleum products must be consistent with reference data as well. This data will exhibit a strong aromatic (alkylbenzene) profile, with weaker alkane and cycloalkane profiles. Naphthalene- and indan-based aromatic compounds will generally be present as well. To identify gasoline the TIC and each profile should be consistent with the patterns obtained from known reference samples with no unexplainable variations.

5.5.2   Distillates and Dearomatized Distillates

Petroleum distillates and their dearomatized analogs are derived from crude oil with considerably less processing than gasoline. Consequently, the composition of these products is more representative of the types and relative proportions of compounds found in crude oil. The primary processing that distillates undergo is fractionation, a separation based on boiling point range. Dearomatized distillates go through additional processing in order to remove aromatic compounds. Distillates and dearomatized distillates are further subdivided based on their boiling point range. Products with a normal carbon range from approximately C_4 to C_9 are deemed light; C_8 to C_{13} qualify as medium, and products starting as low as C_9 are classified as heavy if sufficiently broad in range (at least five consecutive *n*-alkanes). The heavy class also includes narrower range products that start at or above C_{11}. The primary requirement for identification of distillate products is the presence of a normally distributed

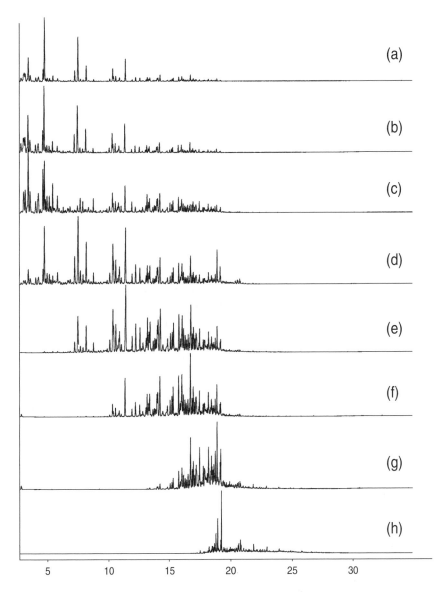

Figure 5.7   Total ion chromatograms of gasoline in various stages of evaporation. (a) Fresh gasoline; (b) 25% evaporated gasoline; (c) 50% evaporated gasoline; (d) 75% evaporated gasoline; (e) 90% evaporated gasoline; (f) 95% evaporated gasoline; (g) 98% evaporated gasoline; (h) gasoline evaporated to dryness.

pattern dominated by normal alkanes. This pattern is readily visible in both the total ion chromatogram and the alkane profile. In addition, a cycloparaffin pattern is expected to be present although at an abundance significantly less than the alkane pattern. The presence of any aromatic compounds will be minimal for a dearomatized distillate; however, a traditional distillate will exhibit recognizable aromatic patterns (benzene-, indan- and naphthalene-based) at abundance significantly lower than that of the alkane pattern, depending on the range of the product. As with the identification of any ignitable liquid, these patterns must correlate to those obtained from a reference liquid of the given class.

5.5.3 Isoparaffinic Products, Normal Alkane Products, and Aromatic Products

Each of these classes of ignitable liquids represents a highly processed product consisting essentially of only a single class of chemical compound. As with distillates, these products may be described further by using the terms light, medium, or heavy to indicate the approximate range of the product.

Isoparaffinic products consist essentially of branched chain alkanes — also known as isoparaffins. Consequently, the patterns obtained for these products will show virtually no aromatic compounds. The pattern of the TIC, alkane profile, and cycloalkane profile will likely show consistent patterns that differ only in abundance. This is due to the fact that isoparaffinic compounds will have spectral peaks that are represented in both the alkane and cycloalkane profiles, although with prudent ion selection these compounds will have a much greater abundance in the alkane profile. Unlike distillates, with a peak distribution that is generally Gaussian or normal in shape, isoparaffinic products tend to have a more sharply defined beginning and end. Identification of peaks will indicate virtually all branched alkanes. There are numerous reference ignitable liquids available for this class of compounds and the patterns obtained for the liquid in question must correlate with those obtained from a known reference liquid.

Analogous to isoparaffinic products are the normal alkane products. Whereas isoparaffinic products contained virtually all branched chain alkanes, normal alkane products contain virtually all straight chain alkanes. There are essentially no aromatic compounds present and any pattern present in the cycloparaffin profile is due to minor ions present in the alkane compounds. The pattern of the TIC, alkane profile, and cycloalkane profile will likely show consistent patterns that differ only in abundance. These patterns tend to be simple, consisting of usually only three to five compounds, although they may have a broader range. Because of the simple nature of these patterns, pattern recognition techniques cannot be used as a sole means for identifying these types of products. It is necessary to also identify the individual constituents by the comparison of retention time and mass spectra to those obtained from known normal alkane references.

In contrast to the alkane-based products just described, aromatic products consist entirely of aromatic compounds and have virtually no aliphatic content. Depending on the boiling point range the composition may be principally benzene based aromatics, or polynuclear aromatics. Regardless of the boiling point range, any patterns present in the alkane or cycloalkane profiles will be minor. The patterns obtained for the various aromatic profiles (benzene-, indan-, and naphthalene-based) must correlate with those obtained from known reference liquids. Additionally, if the pattern is relatively simple, consisting of only a few peaks as with some light aromatic products, it is recommended that the peaks be identified based on retention time and mass spectrum prior to identifying the unknown liquid as an aromatic product.[5]

5.5.4 Naphthenic/Paraffinic Products

Naphthenic/paraffinic products are derived from distillates and consequently retain some of the general features observed in those product types. Like distillates, naphthenic/paraffinic products will have a Gaussian-type distribution with incomplete resolution demonstrated by a rise in the baseline. Unlike distillates, which have a strong presence of normal alkanes, naphthenic/paraffinic products will have very minor contributions from normal alkanes, although branched alkanes and cycloalkanes will be well represented in the pattern. Naphthenic/paraffinic products are dearomatized, and will have little or no aromatic content. The total ion chromatogram of a naphthenic/paraffinic product will appear similar to that of a distillate with a similar boiling point range, if one were able to ignore the spiking normal alkanes that dominate the distillate pattern. The extracted ion profiles obtained for naphthenic/paraffinic products will show virtually no patterns in the various aromatic profiles. The alkane profile will be dominated by branched alkanes and the cycloparaffin/olefin profile will be dominated by cycloparaffins. The overall abundance of the alkane profile and the cycloparaffin profile will be on the same order of magnitude. The patterns present in the TIC and the extracted ion profiles must correspond with those obtained from a known reference source.

5.5.5 Oxygenated Products

Oxygenated products by definition must have at least one significant oxygenated component present. This compound may be present as a single component or may be blended with other compounds, typically hydrocarbons. Identification of an oxygenated product will require identification of the major oxygenated compounds present by a combination of GC retention time and mass spectrum. Other components present may be similarly identified or if the pattern is sufficiently complex, may be identified based on general pattern recognition techniques. It is recommended that these products only be identified when the abundance is sufficient, due to the fact that the partial combustion of ordinary combustibles can create numerous oxygenated compounds. Few generalizations may be made regarding pattern recognition of oxygenated products as most products of this type are the result of blending.

5.5.6   Summary

Identification and classification of ignitable liquids and their residues requires a substantial reference library, a great deal of patience, attention to detail, and significant training and experience in examining ignitable liquids and the typical products of combustion from common fire debris matrices. Because most ignitable liquids are composed of complex mixtures of similar hydrocarbon compounds, the ability to recognize and classify ignitable liquids cannot be based on simple identification of chemical components. It must be based on pattern recognition and comparison with reference ignitable liquids.

5.6   Recent Advances in Instrumental Techniques

While gas chromatography-mass spectrometry continues to be the dominant force in fire debris analysis, research into applications of other instrumental methods continues in an effort to increase sensitivity and specificity, which minimizes the challenges associated with complex chromatograms due to interferences from matrix contributions. One area of research involves the application of gas chromatography-mass spectrometry/mass spectrometry (GC/MS/MS) to fire debris analysis. Recently published work in this area demonstrates that GC/MS/MS may be able to identify ignitable liquids in cases where identification by GC/MS was not possible.[9,10] Research in this area is still in its early stages, and additional studies are needed to demonstrate its advantage over conventional GC/MS techniques.

Another novel approach to the instrumental aspect of fire debris analysis involved a study of the potential utility of comprehensive two-dimensional gas chromatography (GC × GC).[11] The benefit of two-dimensional gas chromatography is that it allows for a much greater resolution due to the fact that the ignitable liquid or extract is subjected to two individual separations, each relying on different characteristics of the components being separated. In this way, GC × GC is better able to separate compounds of interest from those produced by the combustion and pyrolysis of matrix materials. As with the GC/MS/MS method, the potential for better sensitivity in the presence of complex matrices exists primarily due to the enhanced resolution and the subsequent ability to focus more clearly on components of interest. In order to display three dimensions of data (GC × GC × abundance), this technique presents data in a nontraditional format similar to a contour map. Consequently, use of this method will require substantial additional training, even for experienced analysts. However, the improvement in resolution offers optimism in the area of ignitable liquid comparisons, and research into two-dimensional GC methods continues.

Another novel approach to the instrumental analysis of fire debris extracts moves away from chromatographic separations entirely. This research involves using ultra-high resolution mass spectrometry and has shown that some common ignitable liquids can be positively identified without the customary chromatographic separation.[12] Rather than relying on conventional pattern recognition, this method utilizes electron ionization Fourier transform ion cyclotron resonance (EIFT-ICR) mass spectrometry and relies solely on the presence or absence of specific reference compounds within the unknown sample.[12] These compounds are identified by molecular formula (isomers cannot be differentiated) within a broadband spectrum based on high-resolution m/z data. Work in this area has been limited to relatively few ignitable liquids; however, it is expected to continue with emphasis on increasing the number of liquids studied and the number of identifiable reference compounds.[12]

5.7 Conclusion

The analytical methods currently in use for the identification of ignitable liquids and their residues are considerably more sensitive and selective than those previously used. The ready availability of economical benchtop gas chromatograph-mass spectrometers has provided a vast improvement in the quality of analyses that can be performed in a working forensic science laboratory. Innovative methods of data analysis such as target compound chromatography and extracted ion profiling allow the examiner to identify low levels of ignitable liquid residues, even in the presence of overwhelming amounts of unrelated compounds. This ability to separate compounds of interest from matrix contributions has been a major step forward in increasing the sensitivity of current analytical methods, and continues to be the focus of research in the area of ignitable liquid residue analysis.

References

1. Bryce, K.L., Stone, I.C., and Daugherty, K.E., Analysis of fire debris by nuclear magnetic resonance spectroscopy, *J. Forensic Sci.*, 26, 4, 678–685, 1981.

2. Alexander, J., Mashak, G., Kapitan, N., and Siegel, J.A., Fluorescence of petroleum products II. Three-dimensional fluorescence plots of gasolines, *J. Forensic Sci.*, 32, 1, 72–86, 1987.

3. Meal, L., Arson analysis by second derivative ultraviolet spectrometry, *Anal. Chem.*, 58, 4, 834–836, 1986.

4. ASTM E1387-01 Standard Test Method for Ignitable Liquid Residues in Extracts from Fire Debris Samples by Gas Chromatography, ASTM, West Conshohocken, PA, 2001.

5. ASTM E1618-01 Standard Test Method for Ignitable Liquid Residues in Extracts from Fire Debris Samples by Gas Chromatography-Mass Spectrometry, ASTM, West Conshohocken, PA, 2001.

6. Altgelt, K.H. and Gouw, T.H. (Ed.), *Chromatography in Petroleum Analysis*, Marcel Dekker, New York, 1979, pp. 51–69.

7. McNair, H.M. and Miller, J.M., *Basic Gas Chromatography*, John Wiley & Sons, New York, 1997, p. 116.

8. Keto, R.O. and Wineman, P.L., Detection of petroleum-based accelerants in fire debris by target compound gas chromatography/mass spectrometry, *Anal. Chem.*, 63, 18, 1964–1971, 1991.

9. deVos, B.J., Froneman, M., Rohwer, E., and Sutherland, D.A., Detection of petrol (gasoline) in fire debris by gas chromatography/mass spectrometry/mass spectrometry (GC/MS/MS), *J. Forensic Sci.*, JFSCA, 47, 4, 736–756, July 2002.

10. Sutherland, D.A., The analysis of fire debris samples by GC/MS/MS, *Can. Soc. Forensic Sci. J.*, 30, 4, 185–199, 1997.

11. Frysinger, G.S. and Gaines, R.B., Forensic analysis of ignitable liquids in fire debris by comprehensive two-dimensional gas chromatography, *J. Forensic Sci.*, JFSCA, 47, 3, 471–482, May 2002.

12. Rodgers, R.P., Blumer, E.N., Freitas, M.A., and Marshall, A.G., Compositional analysis for identification of arson accelerants by electron ionization fourier transform ion cyclotron resonance high-resolution mass spectrometry, *J. Forensic Sci.*, 46, 2, 268–279, 2001.

ASTM Approach to Fire Debris Analysis

RETA NEWMAN

Contents

0-8493-7885-0/04/$0.00+$1.50

6.1   Introduction

ASTM International is a standards organization that facilitates the development and maintenance of standard practices, methods, and related documents designed to ensure quality work products. ASTM currently has over 130 committees, which maintain over 10,000 standards.[1] One of these committees, the E30 Committee on Forensic Science, is responsible for the development and maintenance of standards associated with forensic analyses and applications. Currently, the forensic discipline that is most comprehensively represented in the ASTM standards is fire debris analysis. These standards are developed and maintained by the Criminalistics Subcommittee E30.01.

The criminal justice community is looking to the forensic science community to produce documents defining minimum criteria for reliable analysis and valid conclusions. ASTM provides an internationally recognized forum for these types of documents. Laboratories that use ASTM standards as the basis of their analysis do so with the consensus of fire debris analysts in the forensic science community supporting their work.

ASTM is a volunteer organization. Anyone interested in developing, maintaining, or voting on forensic science standards may do so by joining ASTM and applying to the E30 Committee. All new and revised standards are balloted to first subcommittee and then main committee members. All ballots are reviewed and all negative votes are addressed. All negative votes must be resolved by the committee prior to acceptance and publishing of the finalized document by ASTM.

Task groups comprised of subject matter experts typically develop ASTM documents. In the late 1980s, representatives from federal, state, local, and private laboratories, concerned about the quality of fire debris analysis and the accuracy of results, formed the original task group to write the initial fire debris analysis standards that were approved and published in 1990. Currently, many of the various scientific and technical working groups including SWG-DRUG (seized drug analysis), SWG-MAT (trace analysis), and SWG-FEX (fire and explosives analysis) serve as subcommittee task groups to develop new standards in their particular forensic specialties.

The E30 committee on forensic science is comprised of 16 subcommittees, the most prolific of which is E30.01 Criminalistics. E30.01 currently has

25 published standards, eight of which are directly related to the analysis of fire debris. The forensic science standards are published in the *ASTM Annual Book of Standards* in Volume 14.02. They can also be found online at http://www.astm.org.

The use of ASTM standards is typically voluntary unless specific government or regulatory agencies require their use. In forensic science the use of ASTM is completely voluntary; however, the use of these documents provides the laboratory, the criminal justice community, and accreditation organizations a means to determine that laboratory practices are valid. Additionally, standard methods, including ASTM standards, do not require extensive validation for use under the ISO 17025 accreditation requirements.[2] While laboratories must always verify that method work is as intended in their individual laboratories, the level of work required is much less extensive than would be required for the validation of a nonstandard method.

ASTM publishes a variety of documents, all of which are called standards. The standards are broken up into different classifications based upon their content and use. ASTM standard documents include classifications, guides, practices, specifications, terminology standards, and test methods. The fire debris standards are currently limited to guides, practices, and test methods. According to the Form and Style for ASTM Standard, a guide is "a compendium of information or series of options that does not recommend a specific course of action." A practice is defined as "a definitive set of instructions for performing one or more specific operations that does not produce a test result." And a test method is "a definitive procedure that produces a test result."[3]

ASTM currently has six standard practices for the separation of ignitable liquid residues from debris. Because these standards do not, by themselves, result in a test result, they are standard practices rather than test methods. Currently standard practices exist for passive headspace sampling with activated charcoal, dynamic headspace sampling with activated charcoal, simple headspace sampling, solvent extraction, passive headspace concentration by SPME, and steam distillation.

There are currently two approved test methods for ignitable liquid analysis. The documents describe the process for instrumental analysis of the fire debris extracts created from the sample separation practices. Specifically, the test methods describe the analysis of fire debris extracts by gas chromatography and gas chromatography-mass spectrometry. They are test methods because they result in data from which conclusions can be drawn.

What follows is an overview of the various standards related to fire debris analysis, including theory, application, advantages, and limitations. This document provides supplemental information and is not intended for, nor does it provide, sufficient detail to substitute for the actual ASTM documents.

6.2   Evidence Handling Standards

6.2.1   ASTM E1492-90 (Reapproved 1999) Standard Practice for Receiving, Documenting, Storing, and Retrieving Evidence in a Forensic Science Laboratory[4]

This document provides direction for proper evidence storage and chain of custody documentation. Although it is not specific to fire debris analysis, the intention is that the document can be applied to any evidence submitted to a forensic laboratory. As a result, the requirements are general in nature. It is, in fact, less rigid than the American Society of Crime Laboratory Directors' Laboratory Accreditation Board (ASCLD/LAB) forensic laboratory accreditation requirements for evidence preservation, packaging, and documentation. If a laboratory meets ASCLD-LAB accreditation requirements for evidence documentation, it will, by default, essentially be in compliance with E1492. This standard was reapproved in 1999, unaltered from its previous revision. It will be up for review and revision in 2004.

6.2.2   ASTM E1459-92 Standard Guide for Physical Evidence Labeling and Related Documentation[5]

Standard E1459 applies to evidence labeling at the scene as well as in the laboratory. It addresses minimum criteria for documenting evidence collection and labeling evidence for subsequent identification. The laboratory criteria include requirements for marking and documenting isolated subitems of evidence separated at the laboratory from original submissions. The requirements of this document are more specific than current ASCLD/LAB laboratory accreditation requirements. Where ASCLD/LAB requires evidence have a unique identifier and handwritten initials of persons sealing the evidence. E1459 specifies that additional information including collection location, date and time of collection, and item descriptions be labeled on each item of evidence.

Unlike E1492, this document is a Standard Guide, rather than a Standard Practice. Thus, its contents are considered recommendations rather than requirements.

6.3   Standard Practices for the Separation of Ignitable Liquid Residues from Fire Debris

The analysis of fire debris is done, almost exclusively, using gas chromatographic (GC) techniques. Sample preparation, thus, necessitates that the ignitable liquid residue be in a vapor or volatile liquid form. Ignitable liquid residue extraction from generally solid and often partially aqueous fire debris is a necessary first step. ASTM E30.01 currently maintains six published standards for the separation of ignitable liquids from fire debris. No single method is appropriate for the analysis of all types of samples or for all classes or ignitable liquids. Each practice has advantages and limitations for use that are specifically addressed in "Scope" and "Significance and Use."

Because method selection can be both sample and case specific, these standards contain information both as to when they are appropriate and when other practices would be more appropriate. Each practice has parameters including temperature, time, volume, etc., that also must be optimized to a given sample or situation. In most cases minimum and maximum parameter ranges are provided. The analyst must then optimize the method within those ranges to meet the needs of the particular sample.

These standard practices were developed and written with the expressed intent that the resultant extracts would be analyzed by either E1387 Standard Test Method for Ignitable Liquid Residues in Extracts from Fire Debris Samples by Gas Chromatography or E1618 Standard Test Method for Ignitable Liquid Residues in Extracts from Fire Debris Samples by Gas Chromatography-Mass Spectrometry.

6.3.1 ASTM E1412-00 Standard Practice for Separation of Ignitable Liquid Residues from Fire Debris Samples by Passive Headspace Concentration with Activated Charcoal[6]

The most commonly used method for separating ignitable liquid residues from fire debris is by adsorption on activated charcoal in passive headspace system. The greatest benefits of this technique include sensitivity and ease of use. While this method generally takes 8 to 16 h to complete, it requires only a few minutes of actual analyst work time. What it lacks in speed, it makes up for in efficiency.

ASTM E1412 is the standard that defines the use, advantages, limitations, and parameters for this technique. This standard was first published in 1991 and revised in 1995. It was extensively updated in 2000 to reflect contemporary research regarding the advantages and limitations of the technique. The name of the standard was changed in 2000 as well. In prior editions, the standard title was ASTM E1412 Standard Practice for the Separation of Ignitable Liquid Residues from Fire Debris by Passive Headspace Concentration. The additional wording "with activated charcoal" was added because E1412 is specific to activated charcoal and because other passive headspace concentration techniques, including SPME and Tenax®, had standards in development.

Passive headspace concentration is the adsorption of headspace vapors in a closed system. Adsorption is the concentration of liquid or vapor (adsorbate) on the surface of a solid (adsorbent). In E1412 the adsorbent is activated charcoal and the absorbate is the volatile compounds vaporized in the head-space of the sample container. It must be noted that the headspace of a fire debris sample typically contains volatile compounds from the matrix, includ-ing pyrolysis and incomplete combustions products, as well as from any ignitable liquid residues that may be present. Thus, this technique, and in fact none of the standard practices described here, provide for exclusive extraction of ignitable liquid residues.

In the simplest, and most common, configuration, activated charcoal (typically impregnated on a polymer strip) is suspended in the headspace of a fire debris container. The container is heated to a constant temperature, typically in an oven; however, in some situations ambient temperatures are appropriate. Volatilized compounds in the headspace that come into contact with the activated charcoal strip become adsorbed. After a period of time, the activated charcoal is removed from the container and washed with a solvent to remove the adsorbed species. The eluant is analyzed using gas chromatographic techniques (Figure 6.1).[7]

Activated charcoal has several advantages as an adsorbent for ignitable liquids. Since most ignitable liquids of interest are petroleum products refined from crude oil, the ideal adsorbent must have an affinity for nonpolar hydro-carbons. Activated charcoal is an excellent nonpolar adsorbent for the col-lection and retention of C6-C20 hydrocarbons.

Unfortunately, all hydrocarbons are not retained equally. Activated char-coal, in fact, most adsorbents, preferentially adhere to different types of absorbates. In the case of activated charcoal, the strength of the adsorption bond (ΔH_{ads}) is roughly two to three times that of condensation (ΔH_{cond}). Less volatile compounds adhere more strongly and thus have longer retention times than more volatile compounds. Additionally, aromatic compounds have a greater affinity for activated charcoal than aliphatic compounds. This can become a complication when the concentration of the adsorbate exceeds the capacity of the adsorbent. A disproportionate representation of higher molecular weight or aromatic compounds — an effect called displacement — can occur. The result will be GC data that disproportionately represents these compounds (Figure 6.2). In most cases displacement does not occur to an extent that would preclude recognition or identification of any ignitable liquids present. The exception would be extremely strong samples; in those instances the adsorption parameters can be adjusted to obtain a more rep-resentative sample.

The key parameters associated with passive headspace concentration are: volatile concentration, system temperature, adsorption time, adsorbent amount, desorption solvent, and desorption solvent volume. All of these parameters are addressed in E1412. Obviously, volatile concentration is not a factor that the analyst can control, however, it is key to determining the

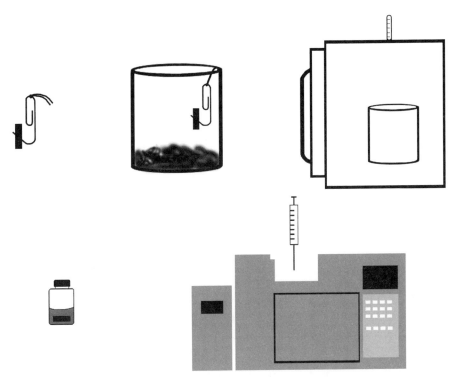

Figure 6.1 Schematic of ASTM E1412-00 passive headspace concentration process using activated charcoal strips. The activated charcoal is suspended in the headspace of the sample container. The container is heated to a constant temperature (usually in an oven). The activated charcoal strip is eluted with a solvent and analyzed by gas chromatographic techniques. (*Source:* Sample Preparation Techniques, *PCFL Fire Debris Analyst Training Manual*, Section 12, Pinellas County Forensic Laboratory, Largo, FL, 2002. With permission.)

appropriate sample-specific values for other parameters, and is so addressed in the standard.

E1412-01 specifies adsorption temperatures in the range of 50 to 80°C for most samples.[6] The system temperature must be high enough to facilitate the vaporization of the higher boiling compounds found in common petroleum products. It also must be low enough to prevent further thermal degradation of the debris and prevent the development of unsafe vapor pressures, especially in the presence of water (steam). It must be noted that even though some ignitable liquid products contain compounds above C18 (octadecane), these compounds may not be sufficiently volatized and sampled in the presence of sorbent matrices, even at elevated temperatures. Charred debris contains active adsorption sites much like activated charcoal. While not nearly as efficient, the adsorbent bonds between higher molecular weight hydrocarbons and debris are often strong enough to preclude volatilization

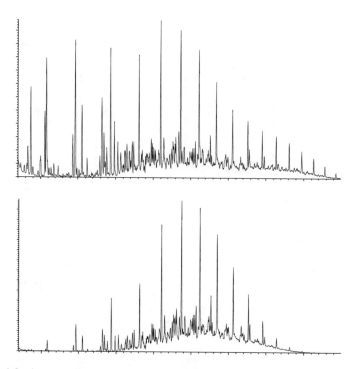

Figure 6.2 Gas chromatographic data illustrating displacement. The top chromatograph represents a neat liquid analysis of a standard accelerant mixture comprised of gasoline, kerosene, and diesel fuel. The bottom chromatogram represents the same liquid collected using an adsorption technique. In this case the amount of ignitable liquid greatly exceeds the capacity of the headspace resulting in displacement — the loss of lower boiling point compounds.

even at higher temperatures.[8] Thus, one key limitation of this technique lies in the inability to distinguish kerosene range (C9 to C18) petroleum products from diesel fuel range (C9 to C23+) petroleum products.

An additional factor in selecting adsorption temperature is that of displacement. Higher temperatures result in faster desorption rates of lower boiling point (i.e., less stronger adsorbed) compounds. Moderate temperatures (60 to 70°C) with extended adsorption times are typical for screening routine samples for the presence of ignitable liquids. Strong samples containing low to medium boiling range ignitable liquids can be often be reliably extracted or reextracted at ambient temperatures.[9]

The adsorption time, i.e., the duration in which the adsorbent is exposed to the headspace, is a function of the system temperature and ignitable liquid concentration. E1412 recommends a general range of 2 to 24 h; 8 to 24 h are typical for sample screening.[9] Like temperature, the adsorption time can be adjusted and the item resampled in the event that the initial data appears over concentrated or highly displaced.

One of the greatest benefits of passive headspace sampling is that it is essentially nondestructive. Most items can be reextracted multiple times without significant loss. Thus, extracts that result in data that appears displaced, distorted, too weak, or too strong can be reextracted using different parameters to achieve better representative samples.

The individual laboratory typically standardizes the amount of charcoal adsorbent used. Although most labs use activated charcoal impregnated on a polymer strip, E1412-01 allows for other methods and sources of activated charcoal including devices created within the laboratory. Because displacement is a function of adsorbent capacity, it is generally recognized that the more adsorbent used, the less displacement will occur. The minimum recommended activated charcoal strip size is defined specified at 100 mm[6] the amount of charcoal for other devices is not specified and thus must be determined and validated by the user laboratory.

The desorption solvent is the liquid used to elute the adsorbed species from the adsorbent to allow for gas chromatographic analysis. The solvents specified by E1412-00 are carbon disulfide, pentane, and diethyl ether. Carbon disulfide is the most efficient resulting in the best representation of the adsorbed species. Both diethyl ether and pentane can result in preferential retention of aromatic compounds on the adsorbent resulting in disproportionate representation of aliphatic compounds in the resultant data.[10] In order of desorption efficiency carbon disulfide performs best, followed by diethyl ether. Pentane is a distant third. Of the three, pentane is, by far, the safest, and thus laboratories must weigh laboratory safety against sample recovery when selecting an elution solvent.

Solvent volume is another parameter defined by this practice. The greater the elution volume, the more efficiently the adsorbed species is removed from the activated charcoal. This must be balanced against the overall loss of sensitivity due to dilution effects and the safety concerns of using larger volumes of dangerous solvents. Volumes in the range of 50 µl to 1 ml are typical, depending on the amount of adsorbent and the choice of solvent.

Quality controls addressed by ASTM E1412-00 include requirements for testing each lot of adsorbents and solvents for contaminants and efficiency. More specific quality controls for the analysis process are addressed in the instrumental test methods rather than in the sample separation practices.

Because this method is essentially nondestructive, both repetitive passive headspace concentration analyses and subsequent alternative separation procedures are possible. This technique is not commonly used for the separation and analysis of compounds that elute prior to pentane or, as described previously, compounds that elute above C18. It is, however, the method of choice as a nondestructive, sensitive technique for sampling C6-C18 compounds from debris headspace.

6.3.2   ASTM E1413-00 Standard Practice for Separation of Ignitable Liquid Residues from Fire Debris Samples by Dynamic Headspace Concentration[11]

Like E1412-00, this standard also addresses the adsorption of volatile compounds from sample headspace on an activated charcoal adsorbent. The difference is that in this process, the system is dynamic rather than passive. The headspace is forced to the adsorbent using a gas or vacuum. As a result, this process is much faster and more sensitive than its passive headspace counterpart. Unfortunately, it does not typically allow for resampling as the volatile compounds are swept from the sample container. Dynamic headspace sampling uses an external force in the form of positive or negative pressure to draw the headspace from the sample container to the adsorbent (Figure 6.3). Because the adsorbent and adsorbate are the same, the parameters are similar to that of passive headspace concentration: temperature, exposure time, adsorbent amount, and elution solvent. Dynamic headspace sampling has the additional consideration of flow, which determines the rate at which the adsorbate comes into contact with the adsorbent.

This standard is divided into two procedures, one for positive pressure apparatus and one for negative pressure apparatus. The positive pressure procedure requires that the laboratory determine the optimal parameters based on an in-house recovery study. The negative pressure apparatus procedure is much more specific in terms of defining the appropriate parameters for temperature, flow, and time. This part of the procedure, as currently written, does not allow for analyst discretion in optimizing extraction parameters. Negative pressure dynamic headspace adsorption is, by far, the more common of the two.

The vapor flow (up to 1500 cc/min)[11] acts to force the contents of the headspace to the adsorbent. Heating the system serves to volatilize ignitable liquid residues. Because the exposure time (5 min) is significantly less than that of passive headspace concentration (up to 24 h), the recommended temperature is somewhat higher (90°C). Temperatures above 90°C are still not recommended due to sample degradation and safety concerns.

The higher kinetic energy of the adsorbate resulting from the flow and elevated temperatures serves to increase the rate of adsorption/desorption of more volatile compounds on the adsorbent. "Breakthrough" is the term given to this dynamic loss of thermally desorbed compounds. Rather than removing a representation of the headspace, dynamic sampling flushes the contents of the headspace from the system. Because the displaced compounds are flushed from the system, they are lost to the sample. Breakthrough is of greater concern to the analyst than simple displacement because representative resampling may not be an option.

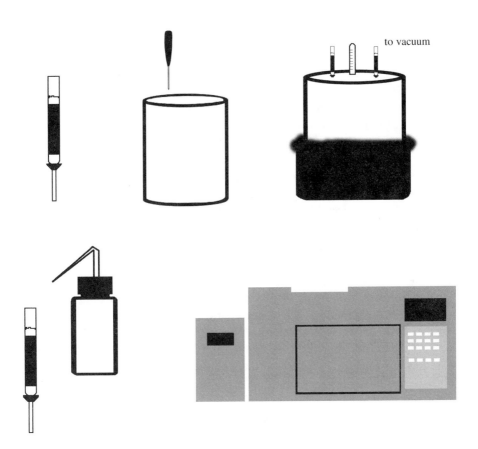

Figure 6.3 Schematic of ASTM E1413-00 dynamic headspace concentration with activated charcoal. Granular charcoal is placed in a glass pipette. Holes are punched into the container to allow for two charcoal-filled pipettes and a thermometer. The container is heated (usually with a heating mantel) and a vacuum is drawn, forcing the headspace to the charcoal collection tube. The second tube serves to filter incoming air. The collection tube is rinsed with a solvent. The eluent is analyzed using chromatographic techniques. (*Source:* Sample Preparation Techniques, *PCFL Fire Debris Analyst Training Manual*, Section 12, Pinellas County Forensic Laboratory, Largo, FL, 2002. With permission.)

Since the advent of activated charcoal strips, fewer laboratories employ dynamic headspace concentration techniques. In the most recent ASTM revision process, there were several motions to have the standard repealed. While this action failed, additional wording was added to the standard regarding the potential destructive nature of the method as well as strong recommendations for sample and extract preservation. The next review/revision of E1413 is scheduled for 2005.

6.3.3 ASTM 1388-00 Standard Practice for Sampling of Headspace Vapors from Fire Debris Samples[12]

Simple headspace sampling of volatile compounds from a closed system headspace is the least sensitive of the headspace sampling techniques. However, it is still used, and generally considered one of the best methods for some ignitable liquid analysis applications. This is a classic technique that dates back to original works in fire debris analysis.

Because it lacks sensitivity, especially for less volatile compounds, this technique is not recommended as a primary technique for the separation and detection of most C6 to C20+ petroleum products. However, it serves as an excellent technique for screening samples for the presence and relative concentration of ignitable liquids and for the analysis of the most volatile compounds of interest to fire debris analysts. The most common application is for the analysis of fire debris for the presence of ignitable liquid components that elute prior to C8 (octane), most notably lower molecular weight, oxygenated compounds (acetone, ethanol, etc.).

This is a very simple process that involves extraction of an aliquot, typically in the range of 0.5 to 2 ml, from the headspace of the sample container for direct injection into the gas chromatograph (Figure 6.4). Depending on the analytes of interest, the container may be heated to facilitate the vaporization of compounds. Like the concentrated headspace sampling techniques, the recommended maximum system temperature is 90°C. Ambient temperatures are often sufficient for the analysis of low molecular weight hydrocarbons and oxygenated compounds.

Headspace concentration techniques, with the possible exception of some SPME applications, are not efficient for the analysis of more volatile ignitable liquid compounds (compounds that elute prior to C6), due to displacement and breakthrough effects. The fact that simple headspace lacks sensitivity is not considered a determent as many of these lower-molecular-weight compounds are formed as incomplete combustion products.[13] Detection at low levels is not typically considered significant. Test Method ASTM 1618-01 for the analysis of ignitable liquid extracts by GC/MS specifies that oxygenated compounds detected at an abundance less than one order of magnitude of that of other matrix compounds should not be considered significant.[14]

ASTM E1388 is simple and nondestructive. Most laboratories include it as an accepted technique for the extraction of low boiling ignitable liquids. Many laboratories use it for sample screening to facilitate the optimization of adsorption parameters for the primary analysis by E1412. This standard was originally published in 1990 and will be up for revision along with most of the other standard practices for fire debris analysis in 2005.

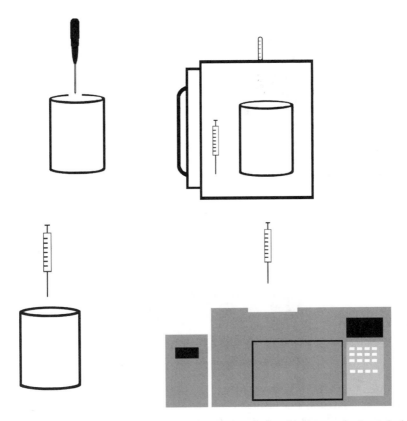

Figure 6.4   Schematic of ASTM E1388-00 simple headspace analysis. A hole is placed in the sample container and covered with tape. The container and the injection syringe are brought to a constant temperature (usually in an oven). An aliquot of the headspace is drawn into the syringe and immediately analyzed using gas chromatographic techniques. (*Source:* Sample Preparation Techniques, *PCFL Fire Debris Analyst Training Manual,* Section 12, Pinellas County Forensic Laboratory, Largo, FL, 2002. With permission.)

6.3.4   ASTM E1386-00 Standard Practice for Separation and Concentration of Ignitable Liquid Residues from Fire Debris Samples by Solvent Extraction[15]

Solvent extraction of fire debris is another technique that most laboratories use on a limited basis. It is not routinely used as a primary technique because it lacks sensitivity, results in messy samples with significant matrix contamination, and is destructive. Solvent extraction involves washing the debris to remove any ignitable liquid components. If necessary, the solvent is concentrated by evaporation prior to instrumental analysis (Figure 6.5).

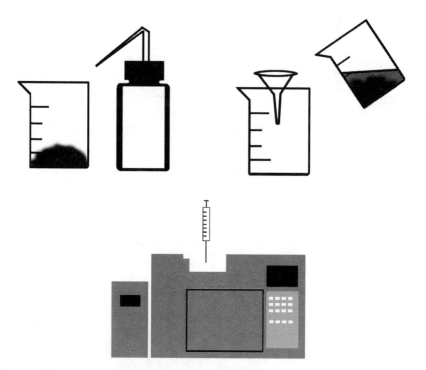

Figure 6.5 Schematic of ASTM E1386-00 solvent extraction. The sample is rinsed with a solvent. The resulting solution is decanted and filtered. The volume is often reduced using nitrogen, air, or low heat. The extract is analyzed by gas chromatographic techniques. (*Source:* Sample Preparation Techniques, *PCFL Fire Debris Analyst Training Manual,* Section 12, Pinellas County Forensic Laboratory, Largo, FL, 2002. With permission.)

The most common solvents are carbon disulfide and pentane. Carbon disulfide is highly toxic but can be purchased relatively pure and has a very low GC/FID detector response. Pentane is much less toxic but can contain impurities. Regardless of the solvent used, common sense and the ASTM standard dictate that each solvent be tested in concentrated form prior to use to detect potential contaminants that could interfere with the analysis.

Solvent extraction is not widely used, but is very advantageous in some situations. The most negative aspect of solvent extraction is that it not only removes volatile components from the debris but can also dissolve other compounds in the substrate. The result is typically a much more complex sample that includes much more matrix interference than would be found in a headspace extraction. The resulting data is typically more complex when synthetic materials are present (i.e., carpeting, pads, foam, etc.). The potential for the substrate components masking the ignitable liquids components is much more likely than any other technique.

On the positive side, this is the only technique that consistently recovers compounds above C18, assuming that matrix interference is minimal. Thus, if the analysis requires differentiation between kerosene range (C9 to C18) and diesel range products (C9 to C23), this is typically the method of choice. Whenever possible, solvent extraction should be employed after concentrated headspace technique due to its destructive nature. Reanalysis using other techniques is not generally possible.

The E1386 standard recommends that, whenever possible, only a representative sample of the debris should be solvent extracted, preserving the rest for any subsequent analysis.[15] Fire debris is not homogeneous, and thus sample splitting is not normally practical. Thus, alternative techniques, usually passive headspace concentration, are recommended for primary analysis. Solvent extraction is then performed as needed. The exception is strong samples or samples with nonporous matrices. In those instances, solvent washing is often the most practical and efficient method for obtaining an extract.

6.3.5 ASTM E2154-01 Standard Practice for Separation and Concentration of Ignitable Liquid Residues from Fire Debris Samples by Passive Headspace Concentration with Solid Phase Microextraction (SPME)[16]

SPME is the newest of the ignitable liquid separation methods. E2154-01 is the first version of the standard. SPME has a number of advantages over the other recovery methods. Most notably it is more sensitive than any other method; it is nondestructive — allowing for multiple analyses, it does not require the use of dangerous solvents, and a single extraction can be completed in a minimal amount of time. Unfortunately, it has several disadvantages, the displacement rate is much higher than that of the activated charcoal methods, extract preservation is not possible, and while the extraction time is minimal, it is less easily automated.

Like E1412, this is an adsorption technique that takes place in a closed system. The adsorbent is a polymer-coated fiber that is contained in a special syringe-like holder. A 100-μm polydimethylsiloxane (PDMS) fiber is recommended for the extraction of hydrocarbons typical of petroleum-based ignitable liquids.[17] The 75-μm Carboxen™/PDMS fiber is generally held to be more effective for the analysis of more polar ignitable liquids like oxygenated solvents.

The apparatus for SPME is more sophisticated than the other adsorption techniques. The fiber is housed in a protective needle on the SPME sampling device. The needle is inserted into the headspace of the container, usually 60 to 80°C, and fiber is exposed by depressing a slide on the sampling device. The headspace is sampled for 5 to 15 min, the fiber is extracted, and the

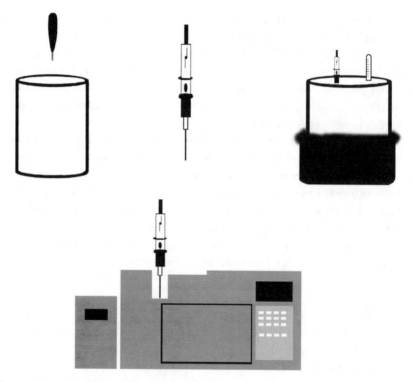

Figure 6.6 Schematic of ASTM E2154-01 solid phase microextraction. A hole is placed in the sample container and secured with tape or septum. The container is heated to a constant temperature either with a heating mantel or in an oven. The SPME holder is inserted through the hole. The plunger is depressed and the SPME fiber is exposed. The fiber is retracted and the extract is analyzed by direct thermal desorption of the fiber in the gas chromatograph. (*Source:* Sample Preparation Techniques, *PCFL Fire Debris Analyst Training Manual,* Section 12, Pinellas County Forensic Laboratory, Largo, FL, 2002. With permission.)

needle removed from the container headspace (Figure 6.6). The fiber is thermally desorbed in the injection port of the gas chromatograph.

The major disadvantage of SPME analysis is that the capacity of the fiber is so small that it is much more susceptible to displacement. In strong samples the displacement can be extreme. Fortunately, since this is a passive technique, multiple analyses are possible, thus the exposure time and/or temperature can be adjusted on samples that appear overloaded or potentially displaced.

The second major disadvantage is the difficulty in "batching" samples for analysis. The SPME fibers are reusable, reportedly up to 100 extracts per fiber. Thus, a blank of the fiber must be run before every extraction to verify that it is free of contaminants. The SPME assembly is manually operated for adsorption and usually manually operated for desorption. Specialized autosamplers are available for SPME; however, they are very costly and not available in most forensic laboratories. Thus, the man-hours required to analyze a ten-item case by SPME is significantly greater than that of the passive activated charcoal method.

SPME has not yet gained widespread use in the fire debris analysis community and, thus, the use of this standard is currently limited as a supplementary or screening technique rather than as a primary technique for ignitable liquid recovery.[16] As more laboratories explore the use of SPME and more validation data (especially in the form of side-by-side analysis or real world samples compared to activated charcoal) is obtained and published, it will likely become an option for a primary method at the next revision scheduled for 2006.

6.3.6 ASTM E1385-00 Standard Practice for Separation and Concentration of Ignitable Liquid Residues from Fire Debris Samples by Steam Distillation[18]

Steam distillation is the least popular, least sensitive, and most complex of the ignitable liquid separation methods. Its use is extremely limited in forensic laboratories and is not intended to be used as a primary extraction technique. Laboratories that use this technique generally do so to obtain a liquid exemplar of ignitable liquids previously identified in the sample by other more sensitive and less destructive techniques. E1385-00 stipulates that the method should only be employed when a portion of the sample can be retained for any subsequent analysis.

Steam distillation is used to separate water from water immiscible liquids. This technique cannot be used to recover water-soluble ignitable liquids. A standard distillation apparatus with a round bottom flask, heating mantle, condensing column, and trap are used (Figure 6.7). A portion of the debris and water are placed in the round bottom flask, and heat is applied to bring the mixture to boiling. Steam is lost through the condensing column and the more volatile hydrocarbons are collected in the trap.[19]

The only advantage of this procedure is that, assuming that the amount of ignitable liquid present is sufficient, a raw liquid is obtained from the extraction. The disadvantages are numerous and include lack of sensitivity, loss of light end components, inefficient recovery of high boiling components, and complex apparatus to include glassware, which must be thoroughly cleaned and must have tight quality control testing.

Like 1413-00 Dynamic Headspace Concentration, this technique was proposed for removal at the last revision. Objections from the few laboratories that use it as a secondary technique resulted in revision to limit use to specific samples, i.e., those with distinctive odors and those that are suitable

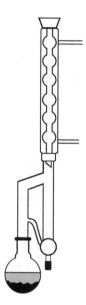

Figure 6.7   Schematic of ASTM E1385-00 steam distillation. Debris is place in a round bottom flask with water. The contents are heated to boiling. The resulting vapor travels to the condensing column where the steam is released from the top and the volatile organic compounds condense into the collection trap. (*Source:* Sample Preparation Techniques, *PCFL Fire Debris Analyst Training Manual,* Section 12, Pinellas County Forensic Laboratory, Largo, FL, 2002. With permission.)

for portioning the sample prior to extraction so that it is not completely destroyed in the analysis.

6.4   Standard Test Methods for the Analysis of Ignitable Liquid Residues in Extracts

There are two standard test methods for the analysis of fire debris extracts. E1387 applies to the analysis of fire debris extracts using gas chromatography with a flame ionization detector (GC/FID) or a gas chromatograph with a mass spectrometer (GC/MS) when the use of the mass spectrometer is limited to producing and evaluating total ion chromatographic data (TIC) only. When the GC/MS is used for extracted ion chromatography or target ion chromatography (as described in Chapter 6), ASTM E1618 is the applicable standard. The standards are very comprehensive and include instrument performance, calibration, and maintenance; sample handling, data analysis, ignitable liquid classification, data interpretation, and reporting. In the most recent revision, these methods were extensively reworded to ensure that, where applicable, the contents were consistent or at least not contradictory.

E1387 was originally published in 1990, along with the original ignitable liquid extraction practices, it was revised in 1994 and 2001. The 2001 version represents the most extensive changes since the first publication. E1618 was originally published in 1994 as a standard guide. In 2001, it became a standard test method. With this change came formal recognition that GC/MS was the recommended technique for ignitable liquid analysis. In the last revision of the GC test method there was a proposal to have E1387 repealed, using the argument that GC/MS was more powerful and resulted in better information for obtaining accurate results.

6.4.1 ASTM E1387-01 Standard Test Method for Ignitable Liquid Residues in Extracts from Fire Debris Samples by Gas Chromatography[20]

Gas chromatography is the traditional method by which fire debris extracts are analyzed. Because ignitable liquids are often comprised of hundreds of compounds, separation is essential to identification or classification. Gas chromatography provides the best means of separating volatile organic compounds, and results in data representing the separation and relative abundance of volatile compounds in a mixture. In the case of fire debris extract analysis; the data represents both ignitable liquid components and volatile compounds produced from the sample matrix.

Data analysis in E1387 is by pattern recognition and pattern comparison to preclassified reference ignitable liquids. Most petroleum based ignitable liquids produce consistent and recognizable chromatographic patterns; these diagnostic peak configurations are used for recognition and classification. The chromatographic properties of the unknown extract are compared to that of known reference ignitable liquids. E1387 provides a general description of each ignitable liquid class and example chromatograms are included. It is not possible to classify ignitable liquids solely on the text and examples provided within the standard.

Gas chromatography is appropriate for routine analysis for most fire debris extracts; however, it has limitations. It does not provide sufficient information for differentiating between some ignitable liquid classes, notably traditional and dearomatized distillates. Additionally, because it does not provide structural information regarding the separated compounds, it cannot be used to directly classify most ignitable liquids nor is it appropriate for the identification of ignitable liquids with few components or without distinctive patterns. Fire debris extracts that result in significant substrate interference (pyrolysis, combustion, incidental compounds), overly complex ignitable liquids, or samples where single or few compound ignitable liquids are indicated should be analyzed using mass spectrometry (E1618).

Laboratories using E1387 as a primary analysis technique should have an extensive ignitable liquid reference collection with numerous exemplars from each ignitable liquid class. The reference ignitable liquids must be classified using an external technique, either by the use of GC/MS (E1618) or by comparison to literature references. Appropriate sources for classification include the Ignitable Liquid Reference Collection Database produced by the Scientific Working Group for Fire and Explosives (SWG-FEX) and managed by the National Center for Forensic Science (NCFS),[21] the *GC/MS Guide for Ignitable Liquids*,[22] or certificates of analysis provided by other organizations that produce or distribute reference ignitable liquids including National Forensic Science Technology Center (NFSTC) and various chemical supply companies. It should be noted, and is specified in E1387 and E1618, that certified reference materials are not required. Reference ignitable liquids obtained through common retail sources are acceptable and are typically one of the best ways to build an extensive collection.

6.4.2 ASTM E1618-01 Standard Test Method for Ignitable Liquid Residues in Extracts from Fire Debris Samples by Gas Chromatography-Mass Spectrometry[14]

E1618-01 provides specific criteria for the use of mass spectrometry for fire debris extract analysis. There are three ways in which GC/MS can be used in ignitable liquid analysis: compound identification/classification, extracted ion chromatography (EIC), and target compound chromatography (TCC). A detailed explanation of the use of GC/MS in fire debris analysis is provided in Chapter 6; an overview will be provided here.

In 1982, Dr. Martin Smith proposed a method of using the data analysis capabilities of a mass spectrometer to extract class indicative ions from a total ion chromatogram.[23] Petroleum based ignitable liquids are comprised of hydrocarbons, specifically alkane and aromatic compounds. Fragmenting these compounds by a mass spectrometer results in common ions that can be extracted from the total ion chromatograph to produce less complex ion chromatograms (Figure 6.8). By extracting chromatograms indicative of the alkane, cycloalkane, mononuclear aromatic, polynuclear aromatic, and indan content of a fire debris extract, more detailed data is generated. This additional level of information can aid in data interpretation by minimizing pyrolysis interferences and allowing the analyst to readily interpret the type and abundance of compounds presents, thus providing key information for ignitable liquid classification.

Ignitable liquid identification by this technique is still made by pattern recognition and comparison. With EIC, more patterns and very specific patterns can be compared. Diagnostic peaks groups that are less abundant and thus difficult or impossible to visualize on the TIC or by GC/FID can

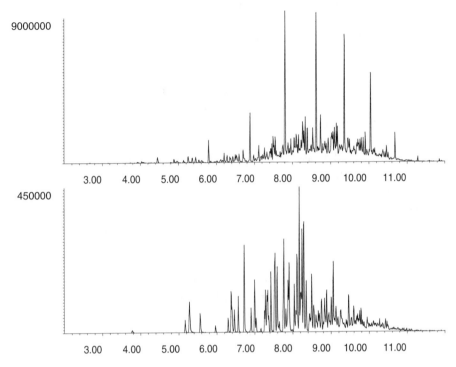

Figure 6.8 Illustration of extracted ion chromatography. The top chromatogram represents the total ion chromatogram of a neat kerosene product. The bottom chromatogram represents the presence and abundance of ions 91, 105, and 119 from the same analysis. The resultant extracted ion chromatogram provides a visual representation of the less abundant, but very important in diagnostics, mononuclear aromatic compounds in the sample.

be easily elucidated using EIC. This method results in data with reduced pyrolysis interference and better visualization of class indicative patterns. It was not until the mid-1990s, when GC/MS became more widely available to the fire debris analyst, that this method became popular. Now, most laboratories use GC/MS with extracted ion chromatography for fire debris extract analysis.

The criterion for use of target compound chromatography (TCC) is also described in E1618. Target compounds that are highly indicative of petroleum products can be isolated, based on retention time and mass spectral fragmentation patterns. Target compound chromatograms are produced by plotting retention time vs. target compound peak abundance.[24] This is not a popular technique; however, it is used a secondary analysis technique by some laboratories. The current standard only provides for target compound analysis for gasoline, medium petroleum distillates, and heavy petroleum distillates.

6.5   ASTM Ignitable Liquid Classification

One of the most important features of E1387 and E1618 is the ignitable liquid classification scheme. This scheme is devised to provide a means of cataloging ignitable liquids, and it provides the forensic science community with a common language to describe ignitable liquid characteristics. The classification tables are identical in current versions E1387 and E1618; however, the data interpretation rules differ significantly in the two documents. This is because GC alone does not provide sufficient information to classify a known or unknown liquid. Classification using E1387 is by comparison to preclassified known ignitable liquids. Mass spectral information, specifically in the form of extracted ion chromatography, does provide sufficient detail to classify ignitable liquids based on chemical composition. It must be noted that identification of an unknown extract is still limited to comparison to a known ignitable liquid.

In the current version, the classification scheme has nine classes of ignitable liquids, most of which are subdivided by boiling range (Table 6.1). Seven of the classes are specific to petroleum-based ignitable liquids: gasoline, distillates, dearomatized distillates, isoparaffinic products, aromatic products, naphthenic paraffinic products, and normal alkane products. One class is reserved for oxygenated solvents and the final class is a catchall miscellaneous class.

The classification scheme has evolved from a numbered six class system to its current appearance which has nine classes subdivided into 25 subclasses. The original scheme was based on a 1980s era ATF ignitable liquid classification. The system was developed from a study of the most common types of ignitable liquids found in debris and commonly available to the public. Those classes were limited to gasoline (Class 2), and light (Class 1), medium (Class 3), kerosene (Class 4), and heavy (Class 5) range petroleum distillates. A miscellaneous class (Class 0) was added to encompass the very few exceptions.

Over time, and as the petroleum industry produced and marketed more elaborate products, the incidence of "miscellaneous" ignitable liquids steadily increased. In 1997, the miscellaneous class of E1618 was expanded to include five subclasses: oxygenated solvents (Class 0.1), isoparaffinic products (Class 0.2), normal alkane products (Class 0.3), aromatic solvents (Class 0.4), and naphthenic-paraffinic products (Class 0.5).[25] E1387-95 was similar but, as it was on a different revision schedule than E1618, it did not include Class 0.5 naphthenic-paraffinic products. To prevent such inconsistencies, in the future, the documents will be revised simultaneously.

In 2001, three separate considerations prompted significant changes to the appearance of the classification scheme. The first was the need to address another type of ignitable liquid — the dearomatized distillate. The second

Table 6.1   ASTM E1387-01 and ASTM E1618-01 Ignitable Liquid Classification Scheme

Class	Light (C_4–C_9)	Medium (C_8–C_{13})	Heavy (C_8–C_{20+})
Gasoline: all brands, including gasohol		Fresh gasoline is typically in the range of C_4–C_{12}	
Petroleum distillates	Petroleum ether Some cigarette lighter fluids Some camping fuels	Some charcoal starters Some paint thinners Some dry cleaning solvents	Kerosene Diesel fuel Some jet fuels Some charcoal starters
Isoparaffinic products	Aviation gasoline Specialty solvents	Some charcoal starters Some paint thinners Some copier toners	Some commercial specialty solvents
Aromatic products	Some paint and varnish removers Some automotive parts cleaners Xylenes, toluene-based products	Some automotive parts cleaners Specialty cleaning solvents Some insecticide vehicles Fuel additives	Some insecticide vehicles Industrial cleaning solvents
Naphthenic-paraffinic products	Cyclohexane based solvents/products	Some charcoal starters Some insecticide vehicles Lamp oils	Some insecticide vehicles Lamp oils Industrial solvents
N-alkane products	Solvents Pentane Hexane Heptane	Some candle oils Copier toners	Some candle oils Carbonless forms Copier toners
Dearomatized distillates	Some camping fuels	Some charcoal starters Some paint thinners	Some charcoal starters Odorless kerosene
Oxygenated solvents	Alcohols Ketones Some lacquer thinners Fuel additives Surface preparation solvents	Some lacquer thinners Some industrial solvents Metal cleaners/gloss removers	
Others, miscellaneous	Single component products Some blended products Some enamel reducers	Turpentine products Some blended products Various specialty products	Some blended products Various specialty solvents

Source: ASTM E1618-01 Standard Test Method for Ignitable Liquid Residues in Extracts from Fire Debris Samples by Gas Chromatography-Mass Spectrometry, *ASTM Annual Book of Standards*, Vol. 14.02, ASTM International, West Conshohocken, PA, 2002. With permission.

was the confusion associated with the ever-expanding class numbering system. The third was the significance of boiling point ranges of ignitable liquids in all classes to both data interpretation and fire investigation.

Dearomatized distillates were not necessarily new to the marketplace; however, they were being found in much greater frequency in common household ignitable liquids, notably charcoal lighter fluids and paint thinners, and thus they were being found more in fire debris. Unfortunately, they did not fit neatly into the descriptions of any of the existing ASTM E1618-97 classes. Much discussion followed as to whether they should be included under the petroleum distillate class (which to date required the presence of aromatic compounds) or the naphthenic-paraffinic class since dearomatized distillates are, in fact, comprised solely of alkanes (paraffins) and cycloalkanes (naphthenes), or whether to place them in their own class. Unfortunately, the predominant pattern characteristics of dearomatized distillates are very different from that of the naphthenic paraffinic products. After much debate it was decided by the task group and agreed to by the subcommittee and main committee to place it in its own class. Cautionary statements were added to E1387 that GC alone might not be sufficient for differentiation of distillates from dearomatized distillates.

The next issue was how to add dearomatized distillates to the classification scheme. Adding another subclass (Class 0.6) under miscellaneous seemed somewhat ridiculous. The numbering system had become a point of confusion rather than a means of clarification. The decision was made to completely eliminate them and use descriptive names for the classes.

The third issue, and one that had an impact on the decision to remove class numbers, was the issue of boiling point. The intended use of ignitable liquids is based on chemical composition and boiling range. The intended product use of a light petroleum distillate is much different than that of a heavy petroleum distillate; this is true of all the ignitable liquid classes. A light isoparaffinic product is common to aviation fuel but would not be seen in a charcoal lighter fluid. Boiling point information can thus be important information in the course of a fire investigation. The decision was made to subclassify all of the ignitable liquid classes, with the exception of gasoline, based on boiling point range.

Another significant change to the classification scheme was the elimination of the kerosene as a boiling point class. Since most laboratories did not differentiate kerosene (Class 4) from heavy (Class 5) petroleum distillates in routine analysis because of the limitations of headspace sampling techniques, the decision was made to eliminate the kerosene range from the classification scheme. Kerosene range products are now included in the heavy product range. The test methods do provide for mechanisms to differentiate products within the range as necessary.

6.6 Conclusion

The ASTM E30 committee on forensic science is very active. New standards representing consensus documents for several areas of forensic science are currently in development. As the judicial system continues to demand external documentation to demonstrate the validity of analytical techniques, it is likely that other areas of forensic science will soon become as well represented in ASTM as fire debris analysis. As membership in ASTM E30 continues to grow and more public laboratories become involved in the ASTM process, the standards will become more abundant and more refined.

While fire debris analysis is the most comprehensively represented in ASTM, there are areas that are yet to be addressed. SWGFEX will be proposing new standards in the near future to address quality assurance, analysis of vegetable oils in fire debris, ignitable liquid extraction using Tenax absorbent, and sample preservation. Technological advancements will likely result in future standards addressing even more sophisticated techniques including GC/GC and GC/MS/MS.

The ASTM fire debris analysis standards have gained widespread use and acceptance in the forensic science community beyond ASTM membership. Critics of the standards are encouraged to express their concerns by joining ASTM and taking an active part in developing and revising the documents.

References

1. *The Handbook on Standardization: A Guide to Understanding Standards Development Today*, ASTM International, West Conshohocken, PA, 2001 (http://www.astm.org/).

2. *ISO/EIC 17025: General Requirements for the Competence of Testing and Calibration Laboratories*, ISO/IEC, Geneva, 1999, p. 12.

3. *ASTM Form and Style*, ASTM International, West Conshohocken, PA, 2003, p. vii (http://www.astm.org/).

4. ASTM E1492-92(1999) Standard Practice for Receiving, Documenting, Storing, and Retrieving Evidence in a Forensic Science Laboratory, *ASTM Annual Book of Standards*, Vol. 14.02, ASTM International, West Conshohocken, PA, 2002.

5. ASTM E1459-92(1998) Standard Guide for Physical Evidence Labeling and Related Documentation, *ASTM Annual Book of Standards*, Vol. 14.02, ASTM International, West Conshohocken, PA, 2002.

6. ASTM E1412-00 Standard Practice for Separation of Ignitable Liquid Residues from Fire Debris Samples by Passive Headspace Concentration With Activated Charcoal, *ASTM Annual Book of Standards*, Vol. 14.02, ASTM International, West Conshohocken, PA, 2002.

7. Dietz, W.R., Improved charcoal packing recovery by passive diffusion, *J. Forensic Sci.*, Vol. 36, p. 111, 1991.

8. Fultz, M.L., Analysis Protocols and Proficiency Testing, *Proceedings of the International Symposium on the Forensic Aspects of Arson Investigations*, pp. 165–194, 1995.

9. Newman, R., Dietz, W., and Lothridge, K., The use of activated charcoal strips for fire debris extraction by passive diffusion. Part I: The effects of time, temperature, strip size and concentration, *J. Forensic Sci.*, 41, 3, 361–370, 1996.

10. Dolan, J. and Newman, R., Solvent Options for the Desorption of Activated Charcoal in Fire Debris Analysis, presented at the American Academy of Forensic Sciences, Seattle, WA, February, 2001.

11. E1413-00 Standard Practice for Separation and Concentration of Ignitable Liquid Residues from Fire Debris Samples by Dynamic Headspace Concentration, *ASTM Annual Book of Standards*, Vol. 14.02, ASTM International, West Conshohocken, PA, 2002.

12. ASTM E1388-00 Standard Practice for Sampling of Headspace Vapors from Fire Debris Samples, *ASTM Annual Book of Standards*, Vol. 14.02, ASTM International, West Conshohocken, PA, 2002.

13. Levin, B., A Summary of the NBS Literature Reviews on the Chemical Nature and Toxicity of the Pyrolysis and Combustion Products from Seven Plastics, National Bureau of Standards, 1986.

14. ASTM E1618-01 Standard Test Method for Ignitable Liquid Residues in Extracts from Fire Debris Samples by Gas Chromatography–Mass Spectrometry, *ASTM Annual Book of Standards*, Vol. 14.02, ASTM International, West Conshohocken, PA, 2002.

15. ASTM E1386-00 Standard Practice for Separation and Concentration of Ignitable Liquid Residues from Fire Debris Samples by Solvent Extraction, *ASTM Annual Book of Standards*, Vol. 14.02, ASTM International, West Conshohocken, PA, 2002.

16. ASTM E2154-01 Standard Practice for Separation and Concentration of Ignitable Liquid Residues from Fire Debris Samples by Passive Headspace Concentration with Solid Phase Microextraction (SPME), *ASTM Annual Book of Standards*, Vol. 14.02, ASTM International, West Conshohocken, PA, 2002.

17. Furton, K.G., Almirall, J.R., and Bruna, J., A simple, inexpensive, rapid, sensitive, and solventless technique for the analysis of accelerants in fire debris based on SPME, *J. High Resolut. Chromatogr.*, 18, 1–5, October 1995.

18. ASTM E1385-00 Standard Practice for Separation and Concentration of Ignitable Liquid Residues from Fire Debris Samples by Steam Distillation, *ASTM Annual Book of Standards*, Vol. 14.02, ASTM International, West Conshohocken, PA, 2002.

19. Brackett, J.W., Separation of flammable material of petroleum origin from evidence submitted in cases involving fires and suspected arson, *J. Crim. Law Criminol. Police Sci.*, 26, 554, 1955–1956, 1995.

20. ASTM E1387-01 Standard Test Method for Ignitable Liquid Residues in Extracts from Fire Debris Samples by Gas Chromatography, *ASTM Annual Book of Standards*, Vol. 14.02, ASTM International, West Conshohocken, PA, 2002.

21. Ignitable Liquid Reference Collection Database, National Center for Forensic Science, Orlando, FL, 2003, (http://www.ncfs.org).

22. Newman, R., Gilbert, M.G., and Lothridge, K., *GC/MS Guide to Ignitable Liquids*, CRC Press, Boca Raton, FL 1998.

23. Smith, R.M., Arson analysis by mass chromatography, *Anal. Chem.*, 54, 13, 1399–1409, 1982.

24. Keto, R.O. and Wineman, P.L., Detection of petroleum-based accelerants in fire debris by target compound gas chromatography-mass spectrometry, *Anal. Chem.*, 63, 1964–1971, 1991.

25. ASTM E1618-97 Standard Test Method for Ignitable Liquid Residues in Extracts from Fire Debris Samples by Gas Chromatography-Mass Spectrometry, *ASTM Annual Book of Standards*, Vol. 14.02, ASTM International, West Conshohocken, PA, 2000.

The Interpretation of Data Generated from Fire Debris Examination: Report Writing and Testimony

7

PERRY M. KOUSSIAFES

Contents

7.1 Interpretation

The interpretation of data generated from fire debris examination usually focuses on recognizing the patterns of ignitable liquids in the chromatograms of the fire debris samples. If fire debris samples contained only ignitable liquids the task of interpretation would be made much simpler and probably could be performed by software alone. However, inherent to fire debris are pyrolysis products that can obscure the ignitable liquid patterns. In Chapter 1 I will attempt to provide an approach to pattern recognition of ignitable liquids and highlight some common features of each class of ignitable liquid (as defined by ATSM), as well as point out some telltale signs of pyrolysis. While pattern recognition software is available and may aide in screening data, ultimately all data must be visually interpreted by an analyst. Once an interpretation has been made, it is important to compare the sample to a known ignitable liquid with a similar pattern.

An excellent source for total ion chromatograms and mass spectra of common ignitable liquids is *GC–MS Guide to Ignitable Liquids* by Newman, Gilbert, and Lothridge. By comparing patterns obtained from the sample to

193

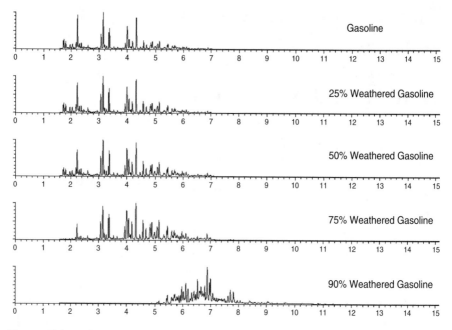

Figure 7.1 Chromatograms of gasoline in various stages of weathering.

patterns of ignitable liquids provided in the book, one can get an idea of the possible ignitable liquid used, if any. It is very important to obtain an aliquot of the ignitable liquid and analyze it on the same instrument that the samples were analyzed on. It may be helpful to replicate the conditions used in the book to run the ignitable liquids to make comparing chromatograms easier. This book is a valuable aid; however, one should not rely on the book exclusively to make identifications.

In general, there are two main factors to consider when attempting pattern interpretation of ignitable liquids, retention time, and target groups. If all the peaks of a chromatogram are clustered in a 4-min window with virtually no peaks outside the cluster, there is a good chance that an ignitable liquid is present. Also, patterns of ignitable liquids may be recognized based on the retention times of the peaks or clusters of peaks. As will be seen, medium and heavy petroleum distillate patterns are visually quite similar. What distinguishes them is the retention time of the alkanes (and, of course, the alkanes themselves). It is also a good idea to have weathered standards available, especially for the more common ignitable liquids such as gasoline. The chromatographic profile of gasoline changes significantly as it weathers. It is sometimes useful to show comparisons of samples to standards to show that it could not be that particular standard. It is not always necessary to show comparisons if the result is determined to be negative or no ignitable liquid is found (see Figure 7.1 and Figure 7.2).

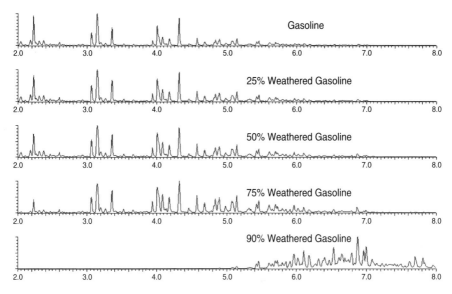

Figure 7.2 Chromatograms of gasoline in various stages of weathering, zooming in on the area of interest.

It is a very good idea to become familiar with the patterns generated by a number of pyrolysis products. This can include burned wood, clothing, carpet, papers, plastics, and building materials. By becoming familiar with common pyrolysis patterns, one can more readily see what is not likely to be an ignitable liquid. Melted plastic bags may have features very similar to kerosene. This is often called "pseudokerosene." These have a picket-fence pattern of alkanes similar to kerosene, but with each alkane there is a corresponding alkene creating a picket fence of doublets or even triplets (two or three peaks grouped close together). These may be classified as "pyrolysis products" or simply as "no ignitable liquid found" (see Figure 7.3).

Anything that may have come in contact with the samples submitted should also be considered. Distinctive patterns may be generated by fire suppressants. Some hand cleaners used by investigators produce a pattern consistent with a medium petroleum distillate.

It is useful to prepare the chromatographic comparisons of samples and standards in a similar fashion each time, according to a set of guidelines or checklist. This makes the task of reviewing data much easier as well as remembering what you did and why, should you have to explain yourself later. It is possible an analyst will no longer be with the lab and you will have to explain his/her work. A checklist like this also insures that nothing is overlooked. For example, if one wishes to make an identification of gasoline, the area from toluene through the dimethylnaphthalenes could be displayed. If mass spectrometry is used, the same sets of spectra should be displayed and evaluated

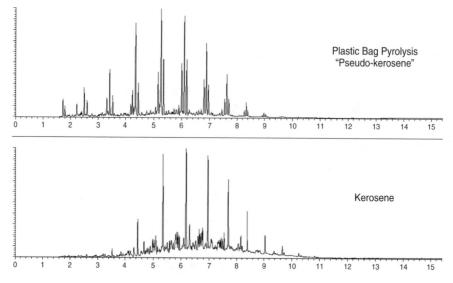

Figure 7.3   A comparison of a pseudokerosene (top) to kerosene (bottom).

each time. One approach is to start with a TIC (total ion chromatogram) to get a total picture of the sample. Then plot SICs (selected ion chromatograms) of selected areas to display specific peaks, with the spectra of the individual peaks following each SIC.

Interpretation is the part of arson analysis that requires the most practice and skill. Unfortunately, no one checklist can state conclusively that if certain components are present, you have a particular ignitable liquid. Only by setting up checklists based on standards run on the same instrument as the samples are run, and by showing comparisons of samples to standards, can a valid conclusion be reached. Be sure to follow the guidelines put forth by ASTM 1387 and 1618 or have a very good reason why you did not follow them.

- Always include a full-scale chromatogram of each sample, in addition to any partial (showing area of interest) chromatograms shown. This will avoid any chance of overlooking a pattern, or of having others suggest that a pattern may have been overlooked.
- If partial chromatograms are shown, make at least one of the partial chromatograms the same for each sample in the case; that is, show the same retention time range for each sample in the case. This will make comparing samples to each other easier. Other retention time ranges for individual samples may then be shown in addition.

It might be helpful for pattern recognition novices to use the following steps to get started in the search for ignitable liquids. Always perform each of the six steps. It is possible that more than one ignitable liquid is present.

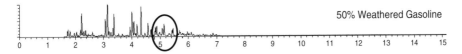

Figure 7.4 Chromatogram of gasoline illustrating CA benzenes.

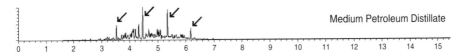

Figure 7.5 Chromatogram of a medium petroleum distillate illustrating alkane series.

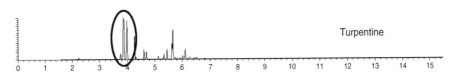

Figure 7.6 Chromatogram of turpentine illustrating terpenes.

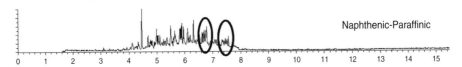

Figure 7.7 Chromatogram of naphthenic-paraffinic product illustrating unusual peak patterns.

1. Look for the C4 benzenes (tetramethylbenzenes, diethylbenzene, Figure 7.4); if found, consider gasoline or aromatics.
2. Look for alkane series (Figure 7.5); if found, consider petroleum distillates, pseudokerosene.
3. Look for terpenes (Figure 7.6); if found, consider presence of wood.
4. Look for early oxygenated compounds:
   - Usually eluted early, often before the solvent
   - Often consist of a single analyte peak
   - May indicate presence of various alcohols or acetone
   If found, consider various alcohols or acetone.
5. Look for unusual peak patterns and clusters (Figure 7.7); if found, consider naphthenic-paraffinic, isoparaffinic, etc.
6. Look for common pyrolysis patterns (Figure 7.8).

The following is a quick guide to interpretation of each of the ASTM classes.

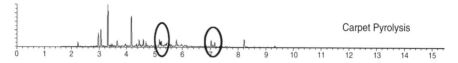

Figure 7.8   Chromatogram of carpet illustrating common pyrolysis patterns.

Gasoline: While not required by ASTM, be able to point out the toluene (m/z 91) and xylenes (m/z 91) that will often be present; trimethylbenzenes (m/z 105), tetramethylbenzenes (m/z 134), indane, and the methylindanes (m/z 117); dodecane (m/z 57), 2- and 1-methylnaphthalene (m/z 142), and the dimethylnaphthalenes (m/z 156). In highly weathered gasoline the early components may not be present. In fresh gasoline the later components, particularly the dimethylnaphthalenes, may be difficult to find.

While not required, the remaining classifications may be further distinguished as light, medium, and heavy ranges. Alkanes are used for purposes of standardizing retention times. For example, the medium range refers to peaks eluting between the retention times of octane and tridecane. Octane and tridecane are not necessarily present.

Light: Most peaks elute before nonane.
Medium: Most peaks elute between octane and tridecane.
Heavy: Most peaks elute after octane with dominant features between decane and eicosane and greater.

Petroleum distillates: A bell-shaped picket fence comprising of at least two consecutive normal alkanes (m/z 57, 71). The alkanes will be dominant features in the chromatogram. Other compounds, often called substrate, may be present at lower levels including aromatics (m/z 91, 105, 134), cycloalkanes (m/z 55, 69), and branched alkanes (m/z 57, 71).

Isoparaffinic products: Abundant branched chain alkanes (m/z 57, 71). No significant normal alkanes, aromatics, or other products. Compounds generally elute in the medium to heavy range.

Aromatic products: Aromatic compounds dominate (m/z 91, 105, 134). No significant alkanes of any type. Compounds generally elute in the light to heavy range.

Naphthenic-paraffinic products: Abundant branched alkanes (m/z 57, 71) and cycloalkanes (m/z 55, 69). No significant aromatics. Compounds generally elute in the medium to heavy range.

Normal alkanes: Normal alkanes (m/z 57, 71) present with no significant substrate of any kind. Compounds generally elute in the medium to heavy range.

Dearomatized distillates: Dominant normal alkanes (m/z 57, 71) with abundant branched alkanes (m/z 57, 71) and cycloalkanes (m/z 55, 69). No significant aromatics. Compounds generally elute in the medium to heavy range.

Oxygenated solvents: Dominant oxygen containing compounds such as alcohols (m/z 31, 45) and acetone (m/z 43, 58). Compounds generally elute in the light range.

Other: Other substances that can be compared to a known ignitable liquid but do not fit any category may be classified as "other."

It is left up to the analyst as to how good a match between samples and standards is required to make the determination that a compound is present or ultimately if it is a particular petroleum distillate. Do not rely exclusively on the computer-generated match factor (if available). It may be that you will wish to use a different classification scheme or that the ASTM classification presented here will be altered. This will require new classification checklists to be made, but the principles of interpretation should remain the same.

Only by setting up your own checklists based on standards you run, and by showing comparisons of samples to standards analyzed on the same instrument, can a valid conclusion be reached. Your retention times (or scan ranges) will likely be different than those shown and will be based on the instrument and conditions used. Retention time ranges need to be updated periodically to account for changes in conditions (column wear, carrier gas flow, etc.). These examples do not cover every possible scenario and you may feel a more complete display of data to be warranted. You may feel a completely different display of data would better serve your needs. It may not be necessary to print spectra for all compounds if the SICs are clear. It may not be necessary to print SICs if the TICs are clear.

See Figure 7.9 through Figure 7.36 for the mass spectral results of some common ignitable liquids.

7.2   Reports

The report should contain the name of the person the analysis is being performed for (typically an investigator), the customer's case number, the lab's case number, the name of the lab, and the name of the analysts who performed the analysis. Of course, the results should be provided. Additionally, a brief explanation of the results may be helpful. This may be on the report itself or provided on a separate sheet of paper. For example, if a medium petroleum distillate was found, one may state that this is a common ignitable liquid often used for charcoal lighter fluids, mineral spirits, and other solvents. If a standard was found that matched particularly well with the sample, this might also be mentioned, but with a disclaimer:

> This sample was consistent with XYZ charcoal starter. This does not necessarily indicate that XYZ charcoal starter was used or that other ignitable liquids could not have been used.

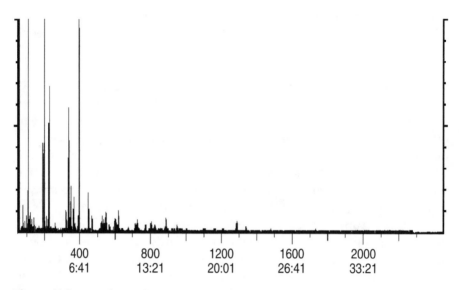

Figure 7.9 Total ion chromatogram of gasoline.

If you suspect the nature of the sample may be responsible for a pattern, state this on the report:

> A heavy petroleum distillate was found. This is consistent with chemicals that might be expected from roof shingles.

This might also be called "negative — expected," indicating that the patterns found were expected so the sample may be considered negative of added ignitable liquid. If needed, a disclaimer regarding the sample container's condition and integrity may be warranted.

> The results may not be valid due to holes in the container of the sample.

or

> The results may have been compromised due to the sample container not being properly sealed for integrity.

A general disclaimer may also be desired on the report. The report is a certificate of analysis, so including things like oven blanks and chromatograms is probably not necessary. These should be included in the case folder and be available in the event they are needed. It may be necessary to issue a report with more than one page. Table 7.1 and Table 7.2 are provided as report examples. Table 7.3 provides a list of ASTM class names and examples.

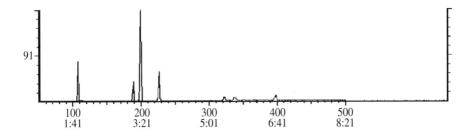

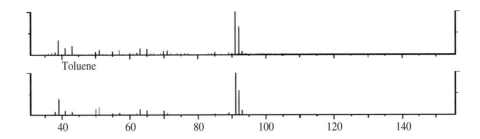

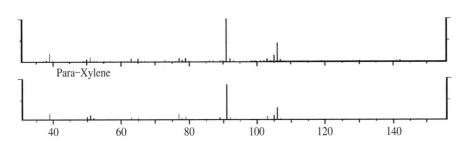

Figure 7.10 to Figure 7.12 Selected ion chromatogram and library-matched mass spectra of toluene and xylenes of gasoline.

7.2.1 Case Folder

The following documents should be kept in the case folder:

- Copy of final report (original sent to investigator)
- Evidence preparation form
- Evidence submission form
- Sample chromatograms
- If it is necessary to expand on a particular area of a chromatogram, place each expansion after the full scale for each sample
- Comparison of samples to standards

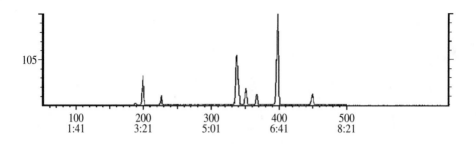

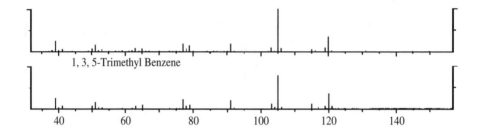

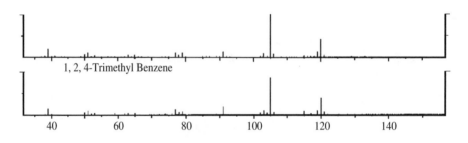

Figure 7.13 to Figure 7.15   Selected ion chromatogram and library-matched mass spectra of trimethylbenzenes of gasoline.

- If it is necessary to expand on a particular area of a comparison, place each comparison expansion after the original scale for each sample
- Sample mass spectra
- Comparison of sample to standard mass spectra
- Instrument blank chromatograms
- This includes instrument conditions
- Chain of custody forms
- Notes page
    - This page is for recording conversations with investigators, attorneys, or for recording any irregularities in the case

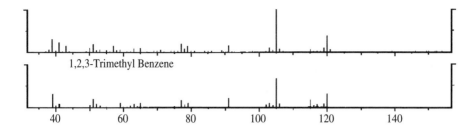

1,2,3-Trimethyl Benzene

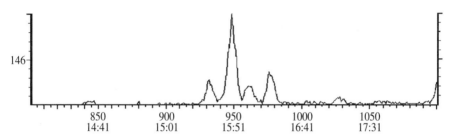

Figure 7.16 and Figure 7.17   Continuation of library-matched mass spectra of trimethylbenzenes.

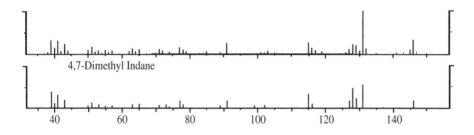

4,7-Dimethyl Indane

Figure 7.18   Selected ion chromatogram and library-matched mass spectra of a dimethylindane of gasoline.

An example of what might be found in a case file of a case containing three samples:

- Final report
- Evidence preparation form
- Evidence submission form
- Sample 1 chromatogram, full scale
- Intensity normalized to tallest peak, full retention time range
- Sample 2 chromatogram, full scale
- Sample 2 chromatogram

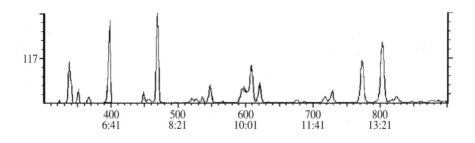

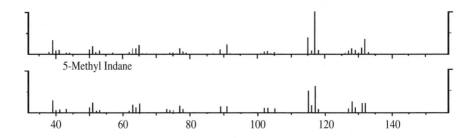

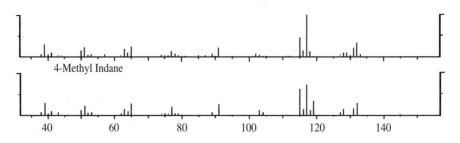

Figure 7.19 to Figure 7.21   Selected ion chromatogram and library-matched mass spectra of methylindanes of gasoline.

- Half intensity scale to better show smaller peaks, full retention time.
- Sample 3 chromatogram, full scale
- Comparison screen showing samples 1, 2, and a gasoline standard
- Intensity adjusted to target compounds, retention time range 2 to 8 min
- Comparison screen showing sample 3 and a kerosene standard
- Intensity adjusted to target compounds, retention time range 2 to 8 min
- Comparison screen showing sample 3 and a kerosene standard
- Intensity adjusted to target compounds, retention time range 4 to 12 min
- Instrument blank chromatogram
- Chain of custody form
- Notes page

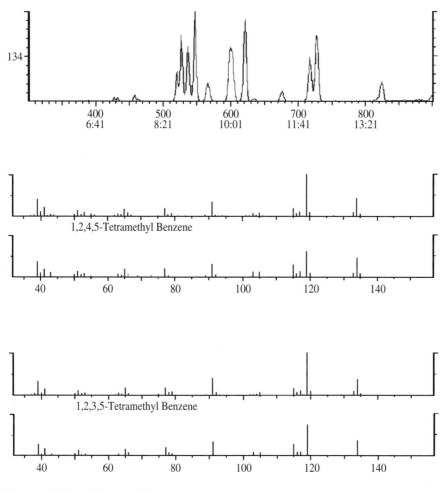

Figure 7.22 to Figure 7.24   Selected ion chromatogram and library-matched mass spectra of tetramethylbenzenes of gasoline. Sometimes m/z 119 is used in place of m/z 134 for identification of tetramethylbenzenes.

   This example did not have any mass spectra. All samples are shown full scale with expansions on areas of interest given as needed. All comparisons are shown at 2 to 8 min with additional comparisons shown as needed. This is done for consistency in the display. If one needs to show the data to a jury, it is much easier to explain the chromatograms when they are all initially on the same scale. QA/QC data is referenced on the evidence preparation forms, but not kept in the case files. This will save file space and still allow the information to be retrieved when necessary. It is helpful to be in the habit of numbering each page in the case file. This will make referring to a partic- ular page much easier should the need arise in court or deposition.

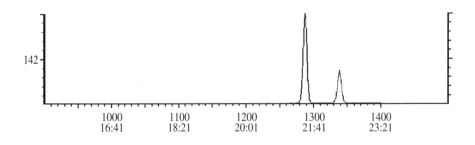

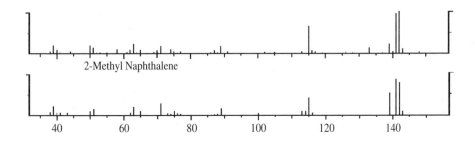

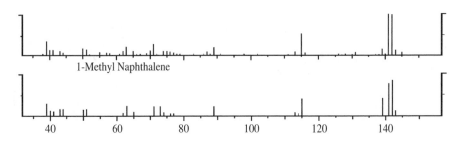

Figure 7.25 to Figure 7.27   Selected ion chromatogram and library-matched mass spectra of methylnaphthalenes of gasoline.

7.2.2   Archival

QA/QC paperwork should be archived and contain the following:

- Initialed and dated run sequence
- Report and chromatogram of test mix
- Chromatogram of oven blank(s)
- Chromatograms of spikes
- Chromatograms of additional test mixes
- Comparison chromatograms of the standards
- The sample extracts should also be archived

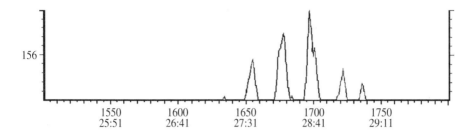

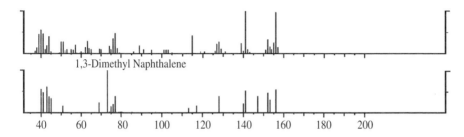

Figure 7.28 and Figure 7.29 Selected ion chromatogram and library-matched mass spectra of dimethylnaphthalenes of gasoline.

The original samples in the sample cans may be returned to the investigator, stored, or, if permissible, disposed of. Photos should be taken of the cans before disposal. These cans are likely to corrode, thus compromising the integrity of the sample. Once this happens, there is no good reason to keep the debris.

7.3 Definitions

The following are definitions of some common terms used in the arson analysis field. It is a good idea to be familiar with them. This list is by no means complete and should be modified as needed.

Accelerant: A substance, often a petroleum-based product that is used to speed up combustion of materials that do not readily burn, e.g., wood. Commonly the object of chemical analysis. Usually has a flash point near or below room temperature. Gasoline is the most common accelerant. Often used interchangeably with ignitable liquid.

Alkanes: Aliphatic hydrocarbons that contain only single bonds. Alkanes are dominant in most petroleum distillates.

Aromatics: A class of chemical substances consisting of alternating double bonds usually derived from benzene. Present in most petroleum distillates. Dominant in gasoline.

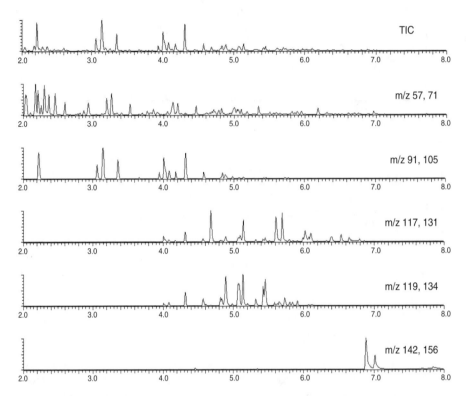

Figure 7.30　An alternate display of gasoline. Retention time range of 2 to 8 min; TIC and SICs shown; individual chromatograms are normalized to the intensity of the tallest peak. This display has the advantage of requiring only one page.

Arson: Definition may vary with jurisdiction. (1) A legal term describing the use of fire to damage property with criminal intent. (2) A legal term referring to a fire set accidentally while in the commission of another crime.

Autoignition temperature: The temperature necessary to produce a flame without an ignition source. Sometimes referred to as spontaneous combustion. *See also ignition temperature.*

Burning: Normal combustion in which the oxidant is molecular oxygen.

Chromatogram: A series of peaks and valleys printed or written on a chart where each peak represents a component or mixture of two or more unresolved components in a mixture separated by gas or liquid chromatography.

Chromatography: A chemical separation procedure which separates compounds according to their boiling point and according to their affinity for an adsorbent or absorbent material.

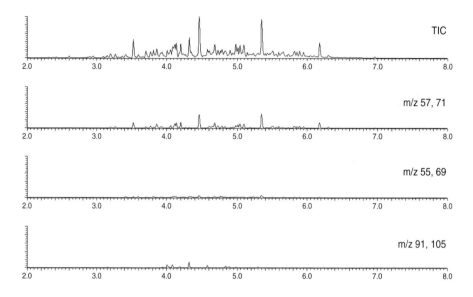

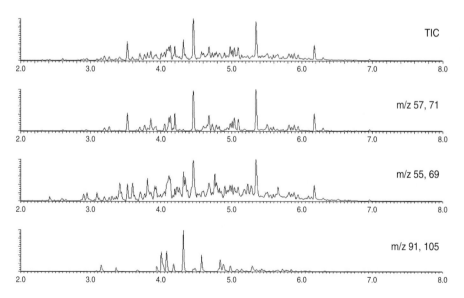

Figure 7.31 A display of the target compounds of a medium petroleum distillate. Retention time range of 2 to 8 min; TIC and SICs shown. The top set of chromatograms has all chromatograms set to the same intensity scale; the bottom set of chromatograms shows each individually normalized.

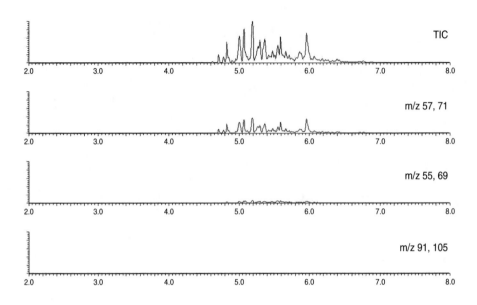

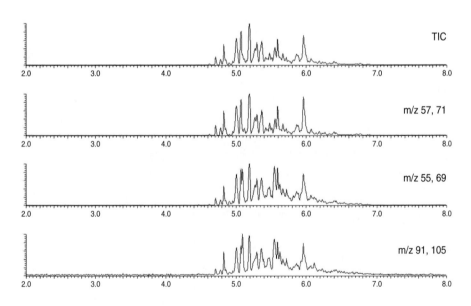

Figure 7.32 A display of the target compounds of an isoparaffinic product. Retention time range of 2 to 8 min; TIC and SICs shown. The top set of chromatograms has all chromatograms set to the same intensity scale; the bottom set of chromatograms shows each individually normalized.

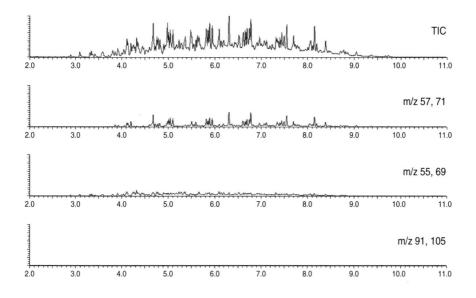

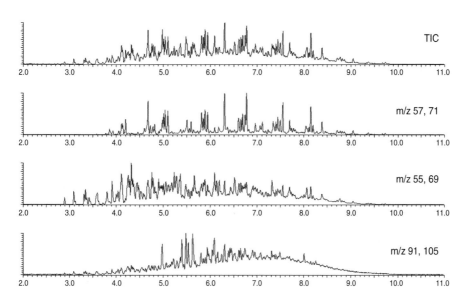

Figure 7.33 A display of the target compounds of a naphthenic-paraffinic product. Retention time range of 2 to 11 min; TIC and SICs shown. The top set of chromatograms has all chromatograms set to the same intensity scale; the bottom set of chromatograms shows each individually normalized.

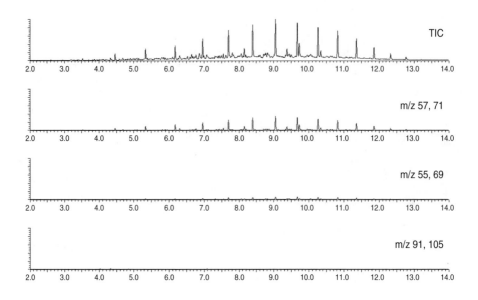

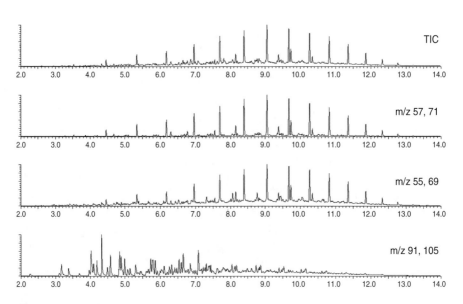

Figure 7.34   A display of the target compounds of a heavy petroleum distillate. Retention time range of 2 to 12 min; TIC and SICs shown. The top set of chromatograms has all chromatograms set to the same intensity scale; the bottom set of chromatograms are each individually normalized.

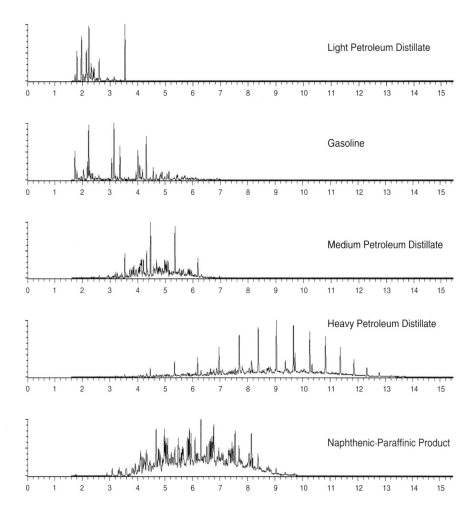

Figure 7.35   Total ion chromatograms of some common ignitable liquids.

Combustible liquid: A liquid that is capable of forming a flammable vapor/air mixture. All flammable liquids are combustible. Whether a liquid is flammable or combustible depends on its flash point and on the agency definition relied upon. The National Fire Protection Association (NFPA) defines a combustible liquid as a liquid having a flash point above 38°C (100°F). Class II, IIIA, and IIB liquids have flash points of 38 to 60°C (100 to 140°F), 60 to 98°C (140 to 208°F), and higher than 90°C (194°F), respectively.

Combustion: Burning; the process by which oxygen combines rapidly with a combustible material in an exothermic reaction. *See fire.*

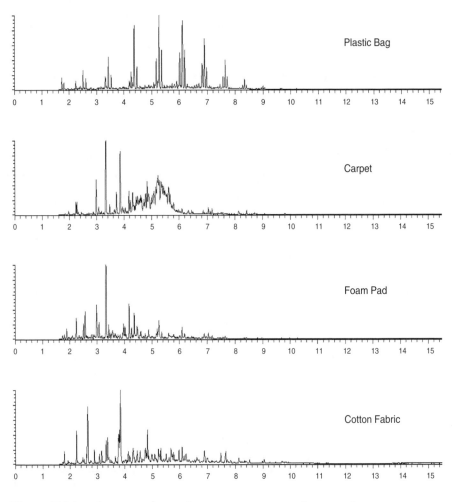

Figure 7.36 Chromatograms of some common pyrolysis products.

Comparison sample: A sample taken which is to be compared to another sample to determine if the same type of ignitable liquid is present in both.

Control sample: A sample taken that is believed not to contain any ignitable liquid. The purpose of this sample is to let the analyst know what compounds are inherent to the sample matrix and thus distinguish them from ignitable liquid in the regular (noncontrol) samples. Not to be confused with a comparison sample.

Cracking: A refining process involving decomposition and recombination of organic compounds (especially hydrocarbons obtained by distillation of petroleum) by means of heat to form molecules suitable for various uses such as motor fuels, solvents, or plastics. Cracking takes place in the absence of oxygen.

Table 7.1   Forensic Laboratory Report

Investigator: Johnson
Miami Police Department
123 Main Street
Miami, FL 33155

Laboratory Number:	020123
Agency Case Number:	02-1234
Receipt Date:	06/04/02
Report Date:	06/07/02

Exhibit Description
  020123-01   SOIL

Results
  A medium petroleum distillate was identified in the following exhibit(s):

  020123-01

  Common examples of medium petroleum distillates include: charcoal lighter fluids, mineral spirits, some wood stains.

The condition of the sample container was acceptable. The integrity of the sample container was acceptable.

The submitted exhibits were examined and interpreted based on a comparison with standards. Quality assurance measures (blanks and spikes) were conducted in association with these exhibits. Should your report contain a result on which you need clarification, please call the Fire and Arson Laboratory.

Subpoenas or correspondence pertaining to this case should refer to the laboratory number.

Perry M. Koussiafes
Crime Laboratory Analyst
Fire and Arson Laboratory
Tallahassee, FL 32300

Cross-contaminate: Allowing the materials present in one sample to come in contact with material in another sample.

Exothermic reaction: A chemical reaction that proceeds spontaneously with the production of heat. All ignitable liquids react by exothermic processes.

Explosion: The sudden conversion of chemical energy into kinetic energy with the release of heat, light, and mechanical shock.

Explosive limit: The highest or lowest concentration of a flammable gas or vapor in air will explode or burn readily when ignited. This limit is usually expressed as volume percent of gas or vapor in air.

Table 7.2   Forensic Laboratory Report

Investigator: Johnson
Miami Police Department
123 Main Street
Miami, FL 33155

Laboratory Number:	020124
Agency Case Number:	02-1235
Receipt Date:	06/04/02
Report Date:	06/11/02

Exhibit Description	**Result**
01 Burned and melted padding	Gasoline

Comments

The condition of the sample container was acceptable. The integrity of the sample container was acceptable.

The submitted exhibits were examined and interpreted based on a comparison with standards. Quality assurance measures (blanks and spikes) were conducted in association with these exhibits. Should your report contain a result on which you need clarification, please call the Fire and Arson Laboratory.

Subpoenas or correspondence pertaining to this case should refer to the laboratory number.

Perry M. Koussiafes
Crime Laboratory Analyst
Fire and Arson Laboratory
Tallahassee, FL 32300

Evaporation: (1) A process in which liquids are transferred into the gas phase. Ignitable liquids undergo evaporation and are eventually lost unless a barrier (container, layer of impenetrable material) can prevent it. (2) A process by which molecules on the surface of a liquid break away and go into the gas phase.

Fire: (1) The light and heat emitted by the rapid oxidation of combustible material. (2) A chemical process where oxidizable material such as hydrocarbons or wood are converted to small and stable molecules such as CO_2, N_2, and H_2O. Accompanied by thermal (heat) and light emission. Requires a fuel, oxidant, and source of ignition. An exothermic reaction. (3) Two types of fires: accidental (a fire that was not deliberately set) and incendiary (a fire that was deliberately set, usually involves the use of an accelerant).

Fire point: The temperature, generally a few degrees above the flash point, at which burning is self-sustaining after removal of the ignition source.

Table 7.3

ASTM Class Name	Examples
Gasoline	Automotive gasoline, all brands and grades including gasohol
Petroleum distillates	
Light	Cigarette lighter fluid, camp fuel, lacquer thinner
Medium	Paint thinner, mineral spirits
Heavy	Kerosene, jet fuel, lamp oil, diesel, fuel oil
Isoparaffinic products	Charcoal starter, copier fluids
Aromatic products	Specialty products
Naphthenic-paraffinic products	Lamp oils, specialty products
Normal alkanes	Lamp oils
Dearomatized distillates	Lamp oils, specialty products
Oxygenated solvents	Liquor, denatured alcohol, rubbing alcohol, fingernail polish remover
Other	Miscellaneous
Negative	No ignitable liquid found
Negative (inconclusive)	The pattern generated does not significantly match available standards or meet criteria established for conclusive determinations
Negative (pyrolysis)	Products normally associated with pyrolysis of the sample matrix
Negative (expected)	Patterns of components consistent with chemicals which are expected to be present

Fire tetrahedron: Fuel, heat, oxygen, chemical chain reaction.

Fire triangle: Fuel, heat, oxygen.

Flame: A rapid gas phase combustion process characterized by self-propagation.

Flame ionization detector: A nearly universal gas chromatographic detector. It responds to nearly all organic compounds. An FID does not respond to nitrogen, hydrogen, helium, oxygen, carbon monoxide, or water. This detector ionizes compounds as they reach the end of the chromatographic column by burning them in an air/hydrogen flame. As the compounds pass through the flame, the conductivity of the flame changes, generating a signal.

Flame point: The temperature at which there is enough vapor to sustain a flame when exposed to an ignition source.

Flammable: Readily burns. The National Fire Protection Association (NFPA) has classified as flammable those liquids with flash points below 38°C and vapor pressures below 275 kPa at 38°C. Within this category, liquids are divided further.

Flashover: A stage in a fire in which all exposed surfaces reach ignition at about the same time.

Flash point: The temperature at which the headspace above a flammable liquid can momentarily produce a flash when exposed to an ignition source. Usually a few degrees below the flame point. Tested with ASTM methods such as the closed cup flash point test. Typical flash points for Class I liquids: gasoline ($-45°C$), class II liquids No. 2 fuel oil ($>38°C$) to No. 5 fuel oil ($>54°C$), class III liquids JP5 fuel ($66°C$) to lube oils and other high boiling fluids.

Forensic science: The application of a scientific discipline to problems having to do with law.

Fuel: A chemical that has the potential of being converted to low energy content molecules such as N_2, CO_2, and H_2O. Releases heat in the process.

Gas chromatography: The separation of organic liquids or gases into discrete compounds seen as peaks on a chromatogram. Separation is done in a column that is enclosed in an oven held at a specific temperature or programmed to change temperature at a reproducible rate. The column separates the compounds according to their affinity for the material inside the column (stationary phase) and their boiling point.

Heat conduction: Transmission of heat through a medium without the perceptible motion of the medium itself.

Heat convection: Heat transfer by fluid motion between regions of unequal density that result from nonuniform heating.

Heat radiation: The propagating waves of heat emitted by radioactivity; consequence of nuclear reaction.

Ignitable liquid: A substance, often a petroleum-based product that is readily ignited when exposed to an ignition source. Commonly the object of chemical analysis. Usually has a flash point near or below room temperature.

Ignition temperature: The temperature necessary to produce a flammable vapor. *See also autoignition temperature.*

Incendiary: Mixtures of oxidizing agents and fuels that can be easily ignited to initiate a fire.

Isopars: A name for a petroleum reformate consisting primarily of branched alkanes. Isopars are devoid of aromatics and usually serve as industrial solvents and copy toners. Some charcoal lighters also fall into this category.

Lowest explosion limit: The lower ratio of air-to-gas mixtures where explosion is possible, e.g., ethyl ether (commonly called ether) will explode in air when present between 2 and 50%.

Mass spectrometry: A method of chemical analysis which vaporizes, then ionizes, the substance to be analyzed, and then accelerates the ions through a magnetic field to separate the ions by molecular weight. Mass spectrometry can result in the exact identification of an unknown compound, and is a very powerful analytical technique, especially when combined with chromatography.

Napthas: A general expression for petroleum distillates, usually in intermediate molecular weight range. The term is poorly defined and should be avoided.

Oxidation: (1) Algebraic increase of oxidation number. Corresponds to the loss, or apparent loss, of electrons. (2) A process where oxygen combines with other elements to generate CO, CO_2, H_2O, and other stable molecules. Other electron transfer reactions are also oxidations. Usually an exothermic (heat-producing) reaction. Fires are a result of oxidation processes.

Paraffin: A mixture of high-molecular-weight alkanes, chemically quite unreactive.

Polyurethane foam: Any of various thermoplastic or thermosetting resins

Pyrolysis: (1) A process where thermal energy (heat) breaks chemical bonds in polymeric materials. The resulting fragments are often volatile. Pyrolysis provides the fuel for matrices that do not undergo unassisted combustion. Wood burns because it pyrolyzes into gas phase volatiles. (2) The breaking down of complex materials into simpler, smaller materials by oxidation or heating. The smaller materials may often recombine, depending on conditions, to make different complex molecules. (3) Something burned.

Saturated hydrocarbon: Compound in which each carbon atom is bonded to four other atoms, and each hydrogen atom is bonded to only one carbon atom. Only single covalent bonds are present.

Spalling: Destruction of a surface, usually concrete, by heat or frost. Due to volume expansion of entrained liquids or volatiles.

Spontaneous combustion: *See autoignition temperature.*

Sublimation: The direct vaporization of a solid by heating without passing through the liquid state.

V pattern: A type of burn pattern that occurs when flames spread upwards and outwards. Often associated with fuel pours along walls.

Vapor: (1) A term describing the gas phase of a liquid or solid. Ignitable liquids can be converted into vapors. Solids such as wood only burn because heat breaks them down into thermal products. (2) Gases formed by boiling or evaporating liquids.

Vapor pressure: The partial pressure of vapor molecules above the surface of a liquid at equilibrium.

Volatility: The ease with which a substance passes from being a solid or a liquid to being a vapor.

Weathering: The evaporation of the more volatile compounds of an ignitable liquid resulting in a greater concentration of the less volatile compounds. May be due to environmental conditions or due to exposure to extreme heat of a fire.

7.4   Additional Background

It is a good idea to have some familiarity with various methods of fire debris analysis. Here is a list of methods of sample preparation that may be obtained from ASTM:

1385   Standard Practice for Separation and Concentration of Ignitable Liquid Residues from Fire Debris Samples by Steam Distillation

1386   Standard Practice for Separation and Concentration of Ignitable Liquid Residues from Fire Debris Samples by Solvent Extraction

1388   Standard Practice for Sampling Headspace Vapors from Fire Debris Samples

1389   Standard Practice for Cleanup of Fire Debris Sample Extracts by Acid Stripping

1412   Standard Practice for Separation and Concentration of Ignitable Liquid Residues from Fire Debris Samples by Passive Headspace Concentration

1413   Standard Practice for Separation and Concentration of Ignitable Liquid Residues from Fire Debris Samples by Dynamic Headspace Concentration

Table 7.4 provides a brief comparison of various sample preparation methods.

7.5   Testimony

Some important guidelines for testifying:

1.   Be objective. *It is not the place of the analyst to convict or exonerate anyone.*
2.   Communicate what the science tells you. *Do not misrepresent the facts nor allow others to misrepresent the facts.*

Table 7.4   Brief Comparison of Various Sample Preparation Methods

Method	Description	Strengths	Weaknesses
Distillation	Sample is distilled in a special apparatus. If the distillate consists of two layers, one or both is injected into the GC	Physical isolation method. Characterization possible by spectral methods (IR, UV) Results in an aliquot of the actual ignitable liquid to show the jury	Very time consuming. Discrimination based on volatility, solubility Lack of sensitivity Destructive method Potential for contamination
Solvent extraction	Sample is extracted with a solvent that is not miscible with water. Resulting solution is injected into GC	Very effective for low volatility substances Polar and nonpolar substances can be recovered All compounds present should be recovered	Significant interference's from matrix Low sensitivity Destructive method Potential for contamination
Direct headspace	Headspace aliquot is withdrawn from the heated debris and is injected into the GC	Very fast, simple technique Minimal opportunity for contamination Nondestructive method	Lack of sensitivity
Passive headspace	Headspace from above the heated sample is adsorbed onto an adsorbent and recovered by solvent or by thermal desorption. Resulting solution is injected into GC	Simple technique Number of samples may be increased with relatively little increase in time required Minimal opportunity for contamination Nondestructive method	Discrimination toward low volatility components
Dynamic headspace (purge and trap)	A gas or air is drawn over a heated sample and adsorbed onto an adsorbent and recovered by a solvent or thermal desorption	Very fast, simple technique Nondestructive method	Cumbersome technique Some potential for contamination Not amenable to multiple sample analysis

Adapted from W. Bertsch, G. Holzer, and C.S. Sellers, *Chemical Analysis for the Arson Investigator and Attorney*, Hüthig, Heidelberg, 1993.

3. Know your own limits of expertise. *If asked a question outside your area of expertise or knowledge, state that you do not know.*
4. Be truthful.
5. Be professional. *A lack of composure may be perceived as a lack of credibility.*

It is quite likely that the attorney you work with will not be familiar with the field of suspect arson analysis. Provided are some questions, along with suggested answers, that you could give to the attorney. It may also be useful to have a list of the instances in which you have been qualified as an expert or have testified in similar cases. *Of course you should change these questions and answers as needed to suit yourself.* One explanation of chromatography is presented. You may be more comfortable with a different but equally valid explanation.

I. QUALIFICATIONS

What is your area of expertise?*
I analyze debris, usually fire debris, for the presence of ignitable liquids. This is often loosely referred to as arson analysis, although I do not determine if the case is, in fact, arson.

Are you an expert at fire investigation in general?
No. All I do is a chemical analysis to determine the presence of ignitable liquids. I do not determine why an ignitable liquid is present, or why it is not present.

Are you an expert on gasoline or other petroleum products?
Only as it applies to analyzing debris. I am not an expert in all phases, uses, or properties of petroleum products.

What education or training do you have that qualifies you to perform arson analysis?*
Degree(s) earned
Relevant short courses
Relevant publications
Relevant experience

Do you have any other professional chemical analysis experience?
Relevant experience

II. STANDARDS USED

What methods or standards do you employ to conduct your chemical analysis of fire debris?
For sample preparation I use ASTM method 1412, a passive headspace concentration method. For analysis and interpretation, I use ASTM methods

1387 and 1618, analysis by GC and GC/MS. These methods are guidelines. Depending on the nature of the sample, I may have occasion to use other methods or to deviate from these methods.

Are these the commonly employed methods in the arson analysis field?
To the best of my knowledge, all labs conducting arson analysis use ASTM methods as guides. I believe that most labs employ the same methods I use.

What is ASTM?
ASTM (the American Society for Testing and Materials) is a not-for-profit organization, one of the largest voluntary standards development systems in the world. They provide standards for testing a wide variety of materials (paints, plastics, textiles, electronics, etc.)

III. TESTING THE SAMPLES

Can you briefly describe what you do to determine the presence of ignitable liquids in samples submitted to you?*
I prepare the samples, and then analyze them on an instrument. Then I interpret the results. My interpretation is based on chemical classifications, not on intended use. Assuming all works well, and it usually does as this is a fairly straightforward procedure using accepted and proven techniques, the interpretation is what requires the skill.

Describe the sample preparation.
This is a lengthy explanation. (*It is a good idea to inform the jury of an upcoming long answer.*)

1. Samples are received in paint cans. These cans never held paint; they were purchased new for the purpose of collecting samples.
2. I inspect the can's condition, look for holes, excessive rust, etc.
3. I observe whether or not the can is evidence-tape sealed. This would verify the integrity of the sample.
4. I look inside the can at the sample and compare what I see to what the investigator wrote on the evidence submission sheet.
5. I spike the sample with a single chemical compound (3-phenyl toluene). This is an easily identifiable compound that does not interfere with the analysis of ignitable liquids.
6. I hang a small strip of carbon inside the can and reseal the can (close the lid).
7. The sample is then baked in a ventilated oven at 66°C, about 150°F. The heat should force any ignitable liquids out of the sample into the

empty space in the can. The carbon strip will then adsorb these, much like a sponge adsorbs water.

8. After a set amount of time (usually 16 h/overnight), the carbon strips are removed from the cans and placed into small vials along with some solvent. The solvent should wash any ignitable liquid from the strip into solution.

9. This vial is then loaded onto an instrument for analysis.

What kind of instrument do you use for analysis?

A gas chromatograph with a flame ionization detector, commonly called a GC or GC/FID. A gas chromatograph/mass spectrometer (GC/MS) may also be used.

How does a gas chromatograph work?

This is another lengthy explanation.

The ignitable liquid, if present, should now be in the solvent. The instrument takes a small amount (less than a drop) of the sample and injects it into the instrument. The sample starts off in an injector port that is very hot. In the injector port the sample is vaporized. It is then carried by a gas into what is called a column. This column is a piece of silica tubing with a coating. This coating helps to separate the sample into its individual compounds. The column is in an oven. This oven is temperature programmed; that is, it is set to start at a relatively low temperature and gradually heated to a higher temperature. So the sample is separated with help from the coating on the column and also by the temperature. The more volatile compounds will exit the column first, and as the temperature is raised, the less volatile compounds will exit the column. The sample is separated by the boiling point of the compounds making up the sample and by the affinity of these compounds for the coating on the column. As the compounds making up the sample exit the column, they pass through a detector. This detector records a signal based on how much compound is passing by. The more compound there is, the stronger or more intense the signal will be. This information, the time the compound passes by the detector relative to when the sample was injected, and the intensity of the signal, is recorded on a graph called a chromatogram. It looks like a series of peaks.

What makes this process useful is that this separation is reproducible. So if I analyze a sample of gasoline today and I analyze a sample of gasoline next week, I should get a similar set of peaks on my chromatogram. That is, similar exit times and similar relative peak intensities. By recognizing the patterns produced I am able to determine what ignitable liquid was present, if any. I do not rely on a single compound, but rather the pattern produced by a series of compounds.

Can you briefly describe the detector?

The FID (flame ionization detector) is a nearly universal gas chromatographic detector. It responds to nearly all organic compounds. An FID does not respond well to nitrogen, hydrogen, helium, oxygen, carbon monoxide, or water. As it turns out, most ignitable liquids are primarily made up of organic compounds. This detector ionizes compounds as they reach the end of the chromatographic column by burning them in an air/hydrogen flame. As the compounds pass through the flame, the conductivity of the flame changes, generating a signal.

Can you briefly describe the MS detector?

This is a method of chemical analysis which vaporizes, then ionizes the substance to be analyzed, and then accelerates the ions through a magnetic field to separate the ions by molecular weight. Mass spectrometry can result in the exact identification of an unknown compound, and is a very powerful analytical technique, especially when combined with chromatography.

A simple way to think about this is that an FID provides two-dimensional data, time vs. intensity. An MS provides three-dimensional data, time vs. intensity vs. ion. For most of the samples analyzed, two-dimensional data is all that is needed. GC/MS is usually used for confirmation when the GC/FID isn't clear enough, as in the case of a complex sample matrix.

IV. QUALITY CONTROL

What quality control measures do you take?

Sample Preparation: I wear gloves during the preparation. I change gloves between each case. This is as much for my safety as well as for preventing contamination.

A new, unused can is treated as a sample and analyzed along with the samples. This is called an "oven blank." The idea is that if any of the sample cans leaked in the oven, they would contaminate the oven blank as well as other samples. If the oven blank is clean, the other samples are probably uncontaminated also.

We also run a spike. This is a can that is spiked with a gasoline/diesel mix. This verifies that the strips do in fact adsorb common ignitable liquids.

Instrumental Analysis: The temperature program is set so that all compounds should be baked out of the instrument. Essentially it is self-cleaning. Between each case I run a blank, different from the oven blank, to check that the instrument is clean. This blank is just the solvent mentioned earlier. This way I know that there could not be any cross-contamination from one case to another. I also run various standards to verify that the instrument is functioning properly.

Report: after I make my interpretation and produce my report, the analyst supervisor reviews my work. If he is not available, another experienced analyst will review my work so that the report can go out. The analyst supervisor will then review my work when he is available. These people are also very experienced in arson analysis. Only after the review can the report be sent to the investigator.

V. IGNITABLE LIQUID IDENTIFICATION

How do you know that what you see is an ignitable liquid, and not just something that might be from the debris?
I analyze various ignitable liquids and also analyze debris in general. The chromatogram produced from burned carpet is very different from the chromatogram produced from paint thinner. Recognizing these differences is where the expertise comes in. It is important to point out that when I identify an ignitable liquid, my identification is made based on chemical classifications, not on intended use.

Can you briefly explain what the different chemical classifications are?
There are eight primary classifications. These distinguish ignitable liquids according to the compounds present and can be further distinguished by boiling point. So a medium petroleum distillate would have compounds with lower boiling points than a heavy petroleum distillate. There is some overlap between the classes. The main exception is gasoline, which is in its own class.
    Sometimes the same liquid may have more than one use. I have seen certain charcoal lighter fluids that were chemically identical to certain mineral spirits. Some are the same, some are different. That's why I only describe my findings based on chemical composition according to chemical classifications. I don't know what the intended use of the ignitable liquid was.
    (*Examples of classes can be provided.*)

Are you able to determine grades or brands of gasoline? That is, if you found gasoline on one sample and gasoline on another sample, could you say for certain that these samples had the same source, came from the same pump?
No, I do not distinguish between grades or brands of gasoline. Different grades of gasoline look pretty much the same to me. I can only identify the sample as being a particular ignitable liquid according to its chemical composition.

Does the composition of an ignitable liquid change during a fire?
Yes. This is commonly called weathering. This refers to the lighter, more volatile components of an ignitable liquid evaporating first, leaving a higher

concentration of the heavier, less volatile components. If the conditions are extreme enough, even these less volatile compounds can evaporate.

Can an ignitable liquid "weather" without being subjected to a fire?
Yes. If left exposed to the environment long enough, a similar weathering pattern will occur. There is no set rule on how long different degrees of weathering will take. This will vary based on the environment and conditions to which the ignitable liquid is exposed. Obviously, when subjected to extreme conditions such as a fire, weathering occurs faster.

VI. THIS CASE (Ask anything; sample question given)

Let's move to this case in particular. Please describe each sample and your findings. Please refer to your notes, if necessary.

Sample 1 was a liquid sample. The evidence submission form from the investigator indicated it is _____. I found _____.

Sample 2 was a wood sample. The evidence submission form from the investigator indicated it is _____. I found _____, etc.

* Questions marked with an asterisk should be asked.

7.6   Spelling List

It is likely that you will be using some words unfamiliar to the court reporter. It is a nice gesture to have a list of these words typed up for him/her. Here are a few:

accelerant	debris	mass spectrometry	phenyl
aliphatic	distillate	meta	phytane
alkane	ethyl	methyl	pristane
alkene	gas chromatography	naphthalene	propyl
analysis	headspace	ortho	pyrolysis
aromatic	helium	para	qualitative
ASTM	hydrogen	paraffin	quantitative
benzene	ignitable	passive headspace extraction	State Fire Marshal
chemistry	isopropanol	Perry Michael Koussiafes	toluene
chromatogram	Koussiafes	petroleum	xylene
chromatography			

New Developments and Quality Assurance in Fire Debris Analysis

8

JOSÉ R. ALMIRALL
JEANNETTE PERR

Contents

8.1   Introduction

Chemical analysis of evidence collected from the scene of a fire presents some unique problems. First, the nature of the sample is not standard, and the sample itself can vary greatly in makeup (type of analytes, type of substrate background, size of sample, and target analyte concentrations). Second, the collection of the sample is also not standard in that there is usually some degradation to the target analytes (the ignitable liquid residues used to accelerate a fire). The safety of individuals and the protection of property takes priority over the preservation of evidence; hence, samples are water soaked or sometimes totally consumed in a fire. The target analytes within the sample are volatile or semivolatile, and therefore the collection process is time sensitive. As shown in the preceding chapters, there are adequate laboratory procedures to detect, identify, and classify ignitable liquid residue (ILR) evidence from the fire scene. The major weakness in the process of evidence analysis during a fire investigation is related to the sampling and the sample preparation step. Finally, it is very straightforward to determine either (1) high quantities of analytes in the presence of a complex and possibly interfering matrix or (2) low levels of analytes in a sample devoid of interfering species.

It is, however, much more difficult to detect and identify low levels of the target analyte species within a complex sample that may include a high background of interfering compounds. It is difficult to determine how many of these samples (containing low levels of analytes characteristic of ignitable liquid residue in a complex matrix) are labeled negative or inconclusive due to the lack of sensitivity of the methods used.

It is also necessary to consider the nature of the sample before an analysis scheme can be recommended. The forensic chemist may be concerned that there are background levels of ILRs in the environment that may be detected in the sample. It is true that some matrices do contain chemical backgrounds that can be confused with ILRs, as they may have similar compositions. There should be, however, a case-by-case consideration of background-level inter-pretation, as this is a sample-dependent characteristic and not a general characteristic of fire scene evidence. For example, one would not expect that a chemical profile of gasoline would be present in a sample taken from a bedroom, in the absence of an alternative explanation such as the sample contained wood that was painted with a mixture diluted with gasoline prior to the application of the paint. In such an unusual case, the interpretation of the evidence would include the analysis of a *control sample* of the wood (taken from a location known not to contain the evidence analyte).

This chapter includes a section on recent developments in sampling and sample preparation, in particular the incorporation of solid-phase microex-traction (SPME) within the sampling and a sample-preparation scheme.[1–11] The section on background and matrix considerations describes some exam-ples of materials that under fire conditions produce compounds that could interfere with fire evidence analysis.[12–17] A section on recent developments in the chemical analysis of fire evidence includes details of developments in GC/MS/MS for this application.[18–24] Finally, a section on quality assurance issues includes some recommendations and current efforts within the exam-iner community of working groups (TWGFEX)[25] and guide/standards writ-ing committees (NIJ, NFPA, ASTM).[26–35]

8.2 New Developments in Extraction Methods

Some of the characteristics of the ideal qualitative analytical method include the following steps: (1) representative (relevant) *sampling*, (2) fast and simple *sample preparation*, (3) efficient *separation*, (4) definitive *compound identifi-cation*, and (5) correct *data interpretation*. The current practice of protocols for fire debris analyses takes a considerable amount of time (up to ~300 h) from Step 1 to Step 5.

Due to the nature of the crime scene sampling and processing of the evidence, and due to the sometimes-considerable case backlogs in operational laboratories, the rate-limiting step is found between Step 1 and Step 2. Another time-consuming step involves the sample preparation.

The sampling of ILRs from fire scene evidence is a challenging task due to the changes that ILRs can undergo prior to the collection of the evidence. Some of the changes include weathering and dilution from evaporation, heat from the fire, and any subsequent fire fighting activities. Additionally, the ILRs within the sample may undergo combustion and pyrolysis. Other minor processes that can hinder the detection of ILRs prior to collection and storage include microbial degradation.[36,37]

The ASTM adsorption–elution method employing passive headspace concentration for the extraction of ILRs from fire scene evidence as described in Chapter 6 uses activated charcoal strips (ACS) to recover the analytes from the sample.[28] A dynamic headspace ASTM method is also sometimes used,[29] followed by elution of the analytes with carbon disulfide or some other suitable solvent.

Solid-phase microextraction (SPME) was first introduced for the extraction of ignitable liquid residues in 1994.[38] SPME of ignitable liquid residues has been shown to be *more sensitive* than ACS extraction due to the elimination of the elution solvent required in the ACS method and also due to the fact that a splitless injection transfers a greater proportion of the extracted analytes onto the column of the GC. Other advantages include a decrease in total analysis time, the elimination of solvents, and the potential for field analysis.[7] SPME has also been favorably compared to headspace, cold trap, and solvent extraction methods and shown to provide several advantages including results with fewer interference peaks.[39] SPME has also been reported to produce *improved selectivity* by decreasing some of the interference problems associated with the analysis of complex matrices (as in the presence of wood or plastic pyrolysis products).[4] More recently, SPME has been reported to improve the sensitivity of extraction of ILRs from fire scene evidence over the ACS method by at least one order of magnitude for most compounds in a standard accelerant mixture (SAM).[40,41]

Solid-phase microextraction is a one-step extraction, preconcentration, and sample introduction method in which a nonvolatile polymeric coating or solid sorbent phase is exposed to a sample. A schematic diagram of the SPME apparatus is shown in Figure 8.1. SPME was named one of the top six ideas in analytical chemistry of the last decade.[42] The analytes can be absorbed or adsorbed onto the nonvolatile polymeric coating or solid sorbent phase, optionally stored, and then desorbed. The desorption process occurs due to heat or the flow of a mobile phase over the fiber, and this heating or flow of a mobile phase over the fiber can present itself in a variety of environments: the injector port of a gas chromatograph (GC), an SPME interface on a high-performance liquid chromatograph (HPLC), or a capillary electrophoresis (CE) instrument. Table 8.1 lists all the commercially available fibers from

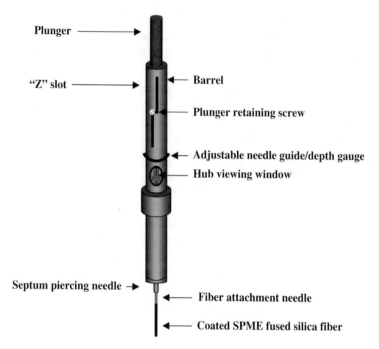

Plunger

"Z" slot

Barrel

Plunger retaining screw

Adjustable needle guide/depth gauge

Hub viewing window

Septum piercing needle →

Fiber attachment needle

Coated SPME fused silica fiber

Figure 8.1   Schematic diagram of SPME apparatus.

Supelco (Bellafonte, PA) with the fibers used for ILR analysis highlighted in bold. The SPME device can be conditioned and reused following the desorption of the analytes. The manufacturer reports that fibers can be reused up to approximately 100 times, although when used in ignitable liquid residue analysis the practical limit is closer to 50 uses or less before the fiber is significantly degraded.

Figure 8.2 represents scanning electron micrographs of six different magnifications of a Carboxen™ polydimethyl siloxane (CAR/PDMS) fiber. The Carboxen particles can be seen in the micrographs. Figure 8.3 represents three different magnifications of a polydimethyl siloxane (PDMS) fiber. Debris and cracks from use can be observed in the micrographs.

CAR/PDMS and PDMS are the two fibers of choice in fire debris analysis. SPME can be used to extract analytes directly from a liquid, a gas, or from headspace samples. In direct liquid emersion, the SPME phase is placed inside the liquid sample. Agitation facilitates the mass transport of the analyte to the fiber for more rapid extraction. The solution immediately surrounding the fiber becomes depleted of the analyte in the absence of stirring or sonication, and the extraction is not optimized.[43]

Since SPME is an equilibrium technique, the headspace extraction follows the equilibrium conditions of the analytes in the sample. That is, there is a bias for the extraction of high-volatility compounds at low temperature

Table 8.1 Experimental Information for Common SPME Fibers

Fiber Coating	Common Abbreviation	Polarity	Max. Temp. (°C)	Operating Temp. (°C)	Conditioning Temp. (°C)	Exposure at Max. Temp. Time (h)	pH	Analytical Application	Extraction Type	Compounds of Interest
Polydimethyl siloxane	**PDMS**	**Nonpolar**	**280**	**200–320**	**250–320**	**0.5**	**2–11**	**GC/HPLC**	**Absorption**	**Volatiles, nonpolar semivolatiles, mid- to nonpolar semivolatiles, ignitable liquid residues**
Polydimethyl siloxane divinylbenzene	PDMS/DVB	Bipolar	270	200–270	250	0.5	2–11	GC/HPLC	Adsorption	Polar volatiles, general purpose, polar volatiles
Polyacrylate	PA	Polar	320	220–310	300	2	2–11	GC/HPLC	Absorption	Polar semivolatiles (specifically phenols)
Carboxen™ polydimethyl siloxane	**CAR/PDMS**	**Bipolar**	**320**	**220–310**	**300**	**1–2**	**2–11**	**GC**	**Adsorption/absorption**	**Gases and volatiles, ignitable liquid residues**
Carbowax® divinylbenzene	CW/DVB	Polar	260	200–250	220	0.5	2–9	GC	Adsorption	Polar analytes (specifically alcohols)
Carbowax® templated resin	CW	Polar	240	No information available	No information available	No information available	2–11	HPLC	Adsorption	Surfactants
Divinylbenzene Carboxen™ polydimethyl siloxane	DVB/CAR/PDMS	Bipolar	270	230–270	270	2–4	2–11	GC	Adsorption/absorption	Odors and flavors

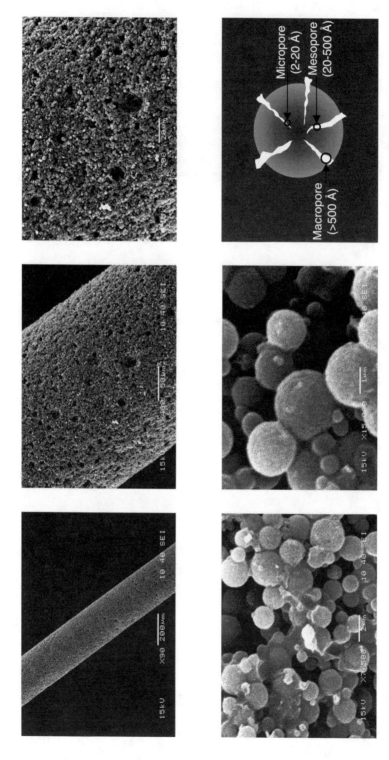

Figure 8.2   Scanning electron micrographs of a Carboxen™ polydimethyl siloxane fiber at different magnifications.

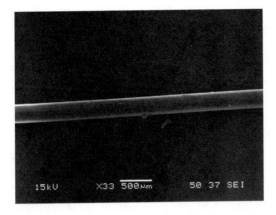

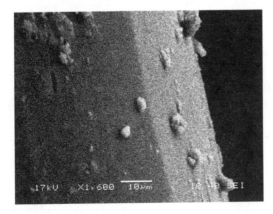

Figure 8.3 Scanning electron micrographs of polydimethyl siloxane fiber at different magnifications.

and a bias for the low-volatility compounds at high temperature. Different fiber chemistries can also affect the equilibrium conditions for a mixture of polar and nonpolar compounds. Headspace extraction at temperatures between 60 and 80°C has been reported for the extraction of ILRs from fire scene evidence. Since SPME is an equilibrium technique and not an exhaustive removal of the analyte from the sample, it has the additional advantage that multiple extractions from the same sample can result in excellent recoveries of the analytes.[4]

The range of compounds extracted by SPME is quite extensive — gases, low molecular weight compounds (mol wt 30 to 225), volatiles (mol wt 60 to 275), alcohol and polar compounds (mol wt 40 to 275), amines, nitroaromatic compounds (mol wt 50 to 300), semivolatile compounds (mol wt 80 to 300), nonpolar semivolatiles (mol wt 80 to 500), nonpolar, high-molecular-weight compounds (mol wt 125 to 600), and surfactants. SPME has also been shown to be useful in the extraction of compounds from complex biological matrices. It was employed to extract the volatiles, methylene chloride, ethanol, and n-alkanes, from the gastric evidence submitted from subjects involved in two different traffic fatalities.[44]

SPME can be configured with different fiber chemistries and in a variety of forms such as coated particles, coated vessels, coated stirring disks, and coated hollow tubes.[45] A commercially available stir bar SPME device (Twister™) made by Gerstel–Global Analytical Solutions (Bremen, Germany) is much more costly than the fiber due to the need to purchase a Thermo Desorption System (TDS) interface into the GC.[46] Fiber SPME devices can be field portable, manual, or automated. The fiber holder can adjust the exposure depth. The field portable holders are equipped with a Teflon™ seal that stores the fiber until the desorption process.

The fiber is very fragile outside of the septum-piercing needle and should be stored within the 23-gauge needle when not in use, immediately following extraction and before sample introduction. The nonbonded SPME phases are stable with some miscible organic solvents, but some swelling of the fiber may occur and the fiber should be allowed to dry before retraction into the needle housing. The nonbonded SPME fiber should not be used to extract from solutions that contain nonpolar organic solvents. Bonded phases are stable with all organic solvents; however, slight swelling may occur with nonpolar solvents. Fiber swelling with headspace extractions (a common form of extraction in fire debris analysis) is very uncommon and should not be a major concern of the forensic chemist. The recommended pH and maximum temperature ranges should be used in order to promote fiber longevity.

During the SPME process, a glass fiber (rod) coated with a small volume of extracting sorbent is exposed to the sample. A common approach is to allow the sample matrix and the fiber coating to reach a partitioning equilibrium, relating the amount of sample extracted on the sorbent to the initial concentration of the analytes in the sample. If the time the sample is exposed to the fiber is insufficient to reach equilibrium, the amount of sample extracted is related to the amount of time sampled. While the ACS (adsorption–elution) method is a three-step sampling/sample preparation/sample introduction technique, the SPME sampling and preconcentration is conducted as a single step for subsequent convenient desorption into the injection of a GC injector.

The equilibrium conditions can be described as Equation 8.1.

$$n = \frac{K_{fs}V_fV_sC_0}{K_{fs}V_f + V_s} \tag{8.1}$$

where n is the number of moles extracted by the coating, K_{fs} is a fiber coating/sample matrix distribution constant, V_f is the coating volume, V_s is the sample volume, and C_0 is the initial concentration of the analyte in the sample. If the sample volume is very large (in comparison to the coating volume), the relationship between the amount extracted on the fiber and the initial concentration of analyte in the sample can be simplified as:

$$n = K_{fs}V_fC_o \tag{8.2}$$

This is of importance to those interested in field sampling as the amount of extracted analyte is independent of sample volume. It is also important to determine the time required to reach equilibrium conditions for a particular sample.

Figure 8.4 shows the effects of extraction time (1 min, 2 min, 5 min, 10 min, and 20 min) on the amount of ignitable liquid component extracted (from a 5-μl spike of a 1000-ppm SAM) using a CAR/PDMS fiber at 70°C.

The time required to reach equilibrium is dependent on the sample, analyte, and extraction temperature conditions used. There is also a direct relationship between the fiber-coating volume and the amount of analyte extracted. As the fiber-coating volume increases, so does the capacity of the fiber and amount extracted. An optimal extraction time needs to be determined for each fiber type and extraction conditions used. Polar and water-soluble analytes (ethanol and light petroleum distillates) are best extracted on a CAR/PDMS fiber at low temperatures. PDMS fibers are used to extract and preconcentrate the medium petroleum distillates and the heavy petroleum distillates (such as diesel fuel) under higher temperature conditions.[10] Figure 8.5 shows the range of ignitable liquid residue target compounds extracted using CAR/PDMS and PDMS fibers.

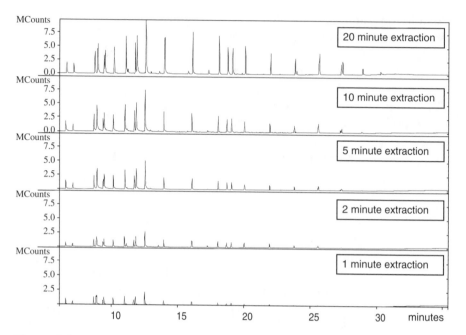

Figure 8.4 Extraction time optimization for CAR/PDMS SPME extraction at 70°C of a 5-μl spike of a 100-ppm SAM.

A recent report on the utility of the SPME method for extracting ILRs indicates some caution as these authors found preferential extraction of the aromatic species using the CAR/PDMS fiber and temperature dependence in the extraction of both aromatic and aliphatic species when the PDMS fiber was used.[47]

SPME was used to identify the presence of gasoline in fire debris evidence where conventional methods, i.e., headspace, were unable to detect the ignitable liquid.[48] Figure 8.6 illustrates the chromatograms resulting from a headspace extraction of a 1-μl spike of a 1000 ppm SAM using a CAR/PDMS SPME fiber (bottom) compared to the extraction of a 1-μl spike of a 1000-ppm SAM using an ACS (8 × 20 mm) eluted with 100 μl of carbon disulfide (top), both plotted on the same scale. The order of the components eluting in the SAM are shown in Table 8.2. Figure 8.7 illustrates the chromatograms resulting from the headspace extraction of a 5-μl spike of a 1000-ppm SAM using a PDMS SPME fiber (bottom) compared to the extraction of a 5-μl spike of a 1000 ppm SAM using an ACS (8 × 20 mm) eluted with 100 μl of carbon disulfide, both plotted on the same scale.

Field portable SPME devices are available from Supelco (Bellafonte, PA) and Field Forensics (St. Petersburg, FL). After extraction using an SPME portable device, the fiber is retracted into the septum-piercing needle and

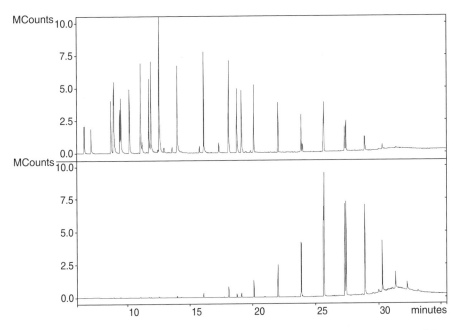

Figure 8.5   Range of ignitable liquid target components in SAM extracted by CAR/PDMS (top) and PDMS (bottom) SPME fibers. (Extraction of a 5-μl spike of a 1000-ppm SAM at 70°C.

Table 8.2   Compounds in Standard Accelerant Mixture (SAM)

n-octane	2-methylnaphthalene
m.p-xylene	n-tetradecane
o-xylene	n-pentadecane
ethylbenzene	n-hexadecane
n-nonane	n-heptadecane
3-ethyltoluene	pristane
1,2,4-trimethylbenzene	n-octadecane
1,2,3-trimethylbenzene	phytane
n-decane	n-nonadecane
cumene	n-eicosane
n-undecane	n-heneicosane
n-dodecane	n-docosane
n-tridecane	n-tricosane
1-methylnaphthalene	

the needle is then housed inside the SPME device where it is sealed with a Teflon cap within the device.

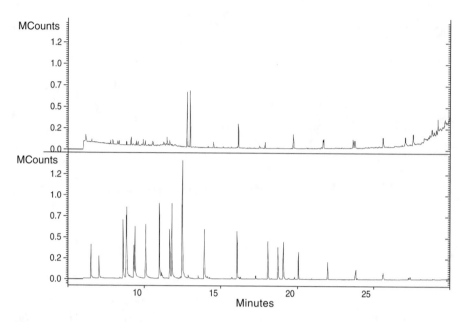

Figure 8.6 CAR/PDMS SPME fiber extraction (bottom) and ACS extraction (top) of a 1-μl spike of a 100-ppm SAM at 70°C.

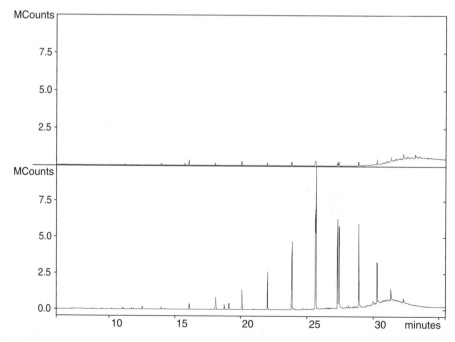

Figure 8.7 PDMS SPME fiber extraction (bottom) and ACS extraction (top) of a 5-μl spike of a 1000-ppm SAM at 70°C.

Despite all the added benefits of SPME, some disadvantages prevent the wide adaptation of the method for routine use in a forensic laboratory. The employment of two different fibers for extraction makes the technique more labor intensive and the technique is more difficult to automate than the liquid injections that result from ACS extractions. The best promise for the technique lies in the development of field sampling methods where rapid screening of samples can take place at the scene of a fire with portable analytical instrumentation.

8.3 Matrix and Background Considerations in Fire Debris Evidence

Heat can be applied to a substance in an uncontrolled manner, as in the case of an accidental fire; in a semicontrolled manner, when a person cooks on a stove; or in a controlled manner, such as an automobile engine. The heat from a fire scene can result in a number of different outcomes depending on the characteristics of the heat and the properties of the substances exposed to the heat. The materials can volatilize, rearrange, or decompose to form many different compounds. The heat causes bonds to absorb energy and become unstable. The instability of the bond causes the bond to break in a certain manner at a specific temperature unique to that type of bond. Most organic compounds become thermally unstable at low temperatures, thereby breaking their bonds homolytically to produce radicals or heterolytically to produce ions. Homolytic bond breakage occurs in radical reactions when each fragment leaves with one bonding electron. Heterolytic bond breakage occurs in polar reactions when one fragment leaves with both of the bonding electrons.[49]

If heating of the substance occurs in air, also known as oxidation, the oxygen reacts with the organic compounds to produce simple products: water, carbon dioxide, carbon monoxide, and further complex products, all known as combustion products. Pyrolysis refers to the thermal fragmentation and degradation of a substance into smaller volatile molecules under heated conditions and in the absence of oxygen or other oxidants. The combustion and pyrolysis of substrate materials present at the scene of the fire can complicate the analysis of an ignitable liquid residue by creating compounds that could interfere with the ILRs. The evidence collected from the scene of a suspect fire usually contains pyrolysis products and combustion products generated during the fire.[50–52] Materials common to our environment can produce compounds through pyrolysis and combustion that are also target components used to classify an ignitable liquid.[12,53,54] Background and pyrolysis products have been characterized from controlled burns of materials found in different environments. The burning of these common materials produce target compounds used in the identification of ignitable liquid residues. These products can interfere with ignitable liquid residue extracts, especially in cases where the quantity of ignitable liquid residue is very small. The sources of interfering compounds include the unburned substrate background products, pyrolysis products resulting from the burning of substrate backgrounds, and combustion products resulting from the burning of substrate background products. Some of the compounds identified as a result of combustion or pyrolysis are also target compounds in ignitable liquid residues. Some compounds used to identify ignitable liquid residues that have also been found in the debris from substrates include tridecane, dodecane, pentadecane, undecane, toluene, naphthalene, *m*-xylene, *p*-xylene, 1-methylnaphthalene, and 2-methylnaphthalene. Figure 8.8 is a chromatogram of compounds detected from the controlled burn of a nylon carpet swatch.[12] Figure 8.9 is a chromatogram of compounds from the controlled burn of paper bag products.[12] A 100-ppm spike of tetrachloro-*m*-xylene (TCMX) was used as an internal standard and added to the elution solvent. A number of target compounds and nontarget compounds can be attributed to the substrate of the evidence.

The presence of background compounds in the environment before the fire[55] and the creation of pyrolysis or combustion products from the burning of substrates can produce complications in the interpretation of the instrumental results.[56] It is very important that a control sample is collected where ILRs are not believed to be present in order to determine what might be related to the background. The results from the analysis of the control sample must be taken into consideration during the interpretation of the results from the analysis of the sample.

8.4 New Developments in Analytical Methods

The current preferred ASTM standards for the analysis and interpretation of ignitable liquid residues are based on separation by gas chromatography and detection using either a flame ionization detector (FID) or a mass spectrometer detector (MS).[34,35] The ILRs are classified as described in Chapter 6. Recent improvements in the analytical selectivity and/or sensitivity for the analysis of ILRs have been reported and will be summarized here. Improvements in sensitivity are necessary in order to detect the extremely small amounts of ILR remaining after a suspected fire. Improvements in selectivity are necessary in order assist in the interpretation to distinguish the ILR from possibly interfering products.

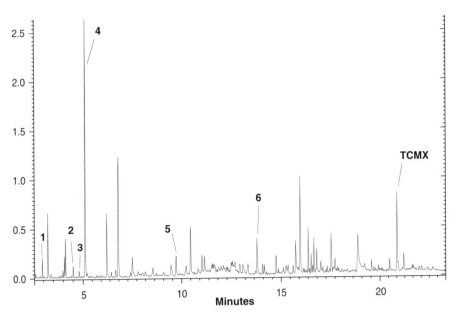

Figure 8.8 Chromatogram of products produced from the controlled burn of a Nylon carpet swatch. Peak identities include: 1 = toluene; 2 = ethylbenzene; 3 = phenylethyne; 4 = styrene; 5 = indene; 6 = naphthalene. (*Source:* J.R. Almirall and K.G. Furton, *J. Anal. Appl. Pyrolysis,* in press; first published online on July 9, 2003. With permission.)

Chromatographic separation for ILRs and pyrolysis/combustion products is typically incomplete, even with lengthy columns (60 m), making for a difficult identification of all the components present. To improve the identification of the eluting peaks, a mass spectrometer is recommended. The mass spectra allow the examiner to identify a peak as a target component based on the molecular ion, if observed, and the fragmentation pattern. An excellent source of reference mass spectra from ILR standards is available.[57] This strategy is effective when the resolution of the chromatography is sufficient to guarantee separation of all the compounds, including any background components present. Forensic scientists rely on patterns of peaks to identify ILRs, but these patterns may be masked if the ILR is present in very small concentrations and the background products are present in high concentrations.

Two-dimensional gas chromatography (GC × GC) was invented in 1991[58] and applied by Philips in 1995[59] to the separation of complex petroleum product mixtures. It is a two-dimensional separation technique that separates all the components of a mixture using two chromatography columns serially connected by a modulator using a short nonpolar column connected to a

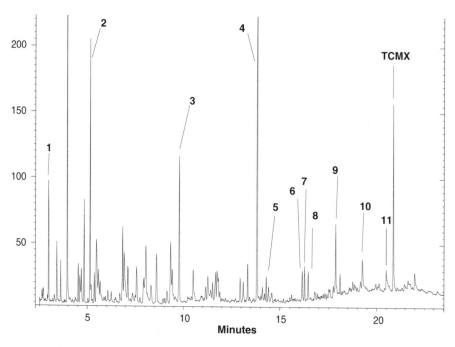

Figure 8.9 Chromatogram of products produced from the controlled burn of paper bag products. Peak identities include: 1 = toluene; 2 = styrene; 3 = indene; 4 = naphthalene; 5 = dodecane; 6 = 2-methylnaphthalene; 7 = tridecane; 8 = 1-methylnaphthalene; 9 = tetradecane; 10 = pentadecane; and 11 = hexadecane. (*Source:* J.R. Almirall and K.G. Furton, *J. Anal. Appl. Pyrolysis*, in press; first published online on July 9, 2003. With permission.)

second polar column. The heater from the modulator periodically rotates over the modulator tube, a thick-filmed section of capillary tubing attaching the two columns together, to desorb trapped analytes and inject them into the second column. Since GC × GC instruments have become commercially available, they have been used for the analysis of complex petroleum samples.[60–62] GC × GC can also be applied to the analysis of fire debris samples in an attempt to identify the presence of ILRs, detect the presence of pyrolysis/combustion products, monitor the changes an ignitable liquid undergoes as a result of weathering, and to determine the differences between ignitable liquids of the same class, i.e., super vs. regular gasoline.[63] By improving the separation of the components of a fire debris sample by GC × GC, it is hoped that the identification of the presence of an ILR will also improve. GC × GC can adequately resolve 1000 components[64] but GC × GC chromatographic programs can take as much as 110 to 130 min. Two-dimensional gas chromatography mass spectrometry (GC × GC/MS) requires a fast mass spectrometer, such as a time-of-flight (TOF) instrument, because narrow second-dimension peaks need to be rapidly sampled in order to obtain a mass spectra. The minimum detected amount (MDA) is lower for GC × GC compared to GC and the minimum concentration detected (MCD) is lower for GC than GC × GC,[65] a disadvantage in the analysis of real casework samples. A second disadvantage is that GC × GC is difficult to optimize. The analysis of simulated fire debris evidence using GC × GC methods has been previously reported.[63]

Electron ionization (EI) is the method of ionization typically used in the mass spectral analysis of ILRs. Electron ionization creates a positive ion by bringing an electron (e−) from a filament held at a potential of −70 eV within close proximity of a molecule causing ionization. The ions survive long enough under low vacuum conditions to be selectively analyzed and detected. Quadrupoles (QMS) and ion traps are the most common types of analyzers but some magnetic sector instruments are still in use.

An ion trap is a mass analyzer that separates ions by applying appropriate electric fields to either store or eject ions according to their mass-to-charge ratio. The ion trap device was described by the German physicists Wolfgang Paul and Helmut Steinwedel in 1953 and patented in 1960.[66] It is commonly referred to as the Paul ion trap, a discovery for which Paul shared a Nobel Prize in Physics in 1989. Figure 8.10 is a photograph of a disassembled ion trap with the major parts labeled. The end caps of the ion trap are either grounded or are biased by an AC or DC potential. The ring electrode is biased by a sinusoidal radio frequency potential. The inner surfaces of electrodes form a three-dimensional quadrupole-like cavity in which ionization, fragmentation, storage, and mass analysis occur. Ion traps are bench-top mass spectrometers that are rugged and easy to disassemble, reassemble, and clean. The ion trap is capable of chemical ionization (CI), electron ionization (EI), selected ion monitoring (SIM), selected reaction monitoring (SRM), multiple reaction monitoring (MRM), and multiple MS experiments (MS^n).[67] Chemical ionization is performed by bleeding a reactant gas into the hyperbolic cavity. The ionization mode can be easily switched back and forth between electron ionization and chemical ionization, even within a single chromatographic run. Ion traps are sensitive due to their fast scan rate, internal ionization source, high-duty cycle, and high efficiency of detection. In comparison with other mass analyzers, they are in the low-to-moderate cost range. Ion traps do not require an extremely high vacuum system (10^{-3} torr vs. 10^{-6} torr for QMS systems).

A thorough theoretical treatment on the ion trap can be found in March and Todd's book,[68] and a recent description of the theory and operation of the ion trap for ILR analysis can be found in a chapter in Yinon's book.[18] This book includes two chapters on the application of GC/MS/MS to the analysis of ILRs.

Figure 8.10   Ion trap with key components labeled.

MS/MS can be achieved by spatially resolving the ions with magnetic sector or quadrupole instruments connected in series or by temporally resolving the ions with the use of an ion trap mass spectrometer. In a triple–quadrupole MS/MS experiment, a second quadrupole acts as a collision chamber to produce collision-induced dissociation (CID) of the selected ions. The third quadrupole, also a mass analyzer, can resolve the resulting daughter ions produced in the collision chamber and detect the ions. MS/MS can be coupled to gas or liquid chromatography systems.

Since the ion trap can store ions for subsequent ejection and detection, an MS/MS experiment with an ion trap will consist of trapping the ion of interest followed by CID and mass analysis within the same space and with the use of scan functions.

One approach to MS/MS for ILR analysis using the internal ionization ion trap is to create the ions of interest by sending a pulse of electrons into the trap and selecting one (or more) ion (called parent or precursor ion) and subjecting that ion to CID by colliding with an inert gas such as helium. This precursor ion should be characteristic of the analyte molecule and, preferably, the parent ion of the molecule. The advantage of an MS/MS experiment for the analysis of ILRs is the ability to isolate target components from coeluting peaks of interfering species. Both the selectivity and sensitivity of the detection is improved. In aromatic compounds the molecular ion is selected, while in aliphatic compounds a typical fragment ion such as m/z 85 is selected.

Figure 8.11 illustrates the chromatogram of pyrolysis products (top) and pyrolysis products with a number of the components in gasoline added to form a total mixture concentration of 1000 ppm as a gasoline standard (bottom) using GC/MS analysis. Burning a small section of nylon carpet inside a paint can and depriving the burning material of oxygen formed the pyrolysis products. Using the ACS method and eluting the ACS with 100 µl of carbon disulfide extracted the products. Figure 8.12 illustrates the same

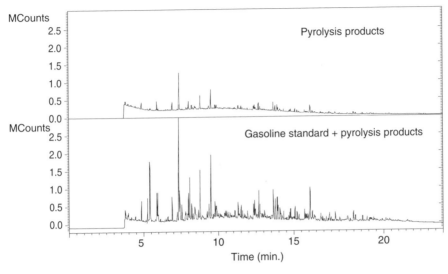

Figure 8.11   The chromatogram of pyrolysis products alone (top) and pyrolysis products with a spike of gasoline added to form a total mixture concentration of 1000 ppm of gasoline (bottom), using GC/MS analysis.

samples analyzed under the GC/MS/MS conditions. The GC/MS/MS chromatograms are more easily interpreted and the gasoline components (bottom) are clearly isolated from the coeluting interfering compounds from the burned carpet. The retention times for the target compounds must be known and the MS/MS conditions must be predetermined for each analyte compound of interest. A gasoline standard can be used to determine the retention times and MS/MS conditions.

GC/MS/MS provides improved selectivity for the identification of target compounds when coeluting compounds mask the compounds of interest. The detection limit (S/N>3) for a single-component target compound in an ILR is ~10 times lower when using GC/MS/MS in the splitless mode over GC/MS in the splitless mode. The combination of improved selectivity and improved sensitivity provides GC/MS/MS with the potential for improving the *detection* and *identification* of target compounds in residues extracted from fire debris, especially in cases where the sample concentration is very low and/or when the sample contains interfering species. It is expected that future work will include the generation of compound-specific MS/MS spectra under standardized conditions for target compounds of interest in fire debris analysis. Additional examples and other approaches to the application of MS/MS for the analysis of ILRs in fire debris have been reported.[18–24] GC/MS/MS is a simple to use and highly discriminative tool to analyze ILRs and should gain wider application amongst the forensic community after further exposure and validation of this method.

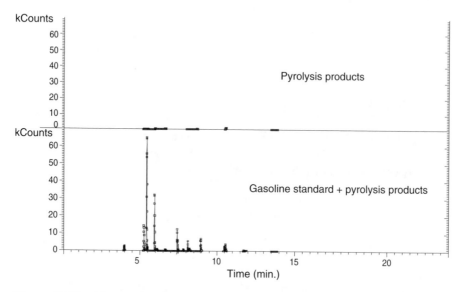

Figure 8.12 The GC/MS/MS analysis of pyrolysis products alone (top) and pyrolysis products with gasoline added to form a total mixture concentration of 100 ppm of gasoline (bottom).

Another recent development has been the reports of the application of Fourier transform ion cyclotron resonance mass spectrometry (FT ICR/MS) to the analysis of ILRs. FT ICR/MS can trace its beginnings back to the late 1940s.[69,70] FT ICR/MS offers ultra high resolution, high mass accuracy, the baseline resolution of multiple species at the same nominal mass without prior chromatographic separation, and rapid analysis.[71] The capabilities of this method include the assignment of a molecular formula for each peak in the mass spectrum producing an "elemental fingerprint" of an ignitable liquid and weathered ignitable liquid.[72] The resulting mass spectra, however, are graphically complex. FT ICR/MS instruments are very expensive and require an extremely high vacuum. Recent reports of the applications of this method for the analysis of ILRs are very promising, but forensic laboratories may find the complexity and cost of the method to be a significant obstacle to the adoption of this method as a routine tool.

8.5 New Developments in Data Analysis

The data generated from ignitable liquid analysis has recently been manipulated either by individual extracted ion profiling[73] or with multivariate pattern recognition.[74] Extracted ion profiling can assist the analyst in distinguishing an ignitable liquid residue from interfering compounds by selecting ions that are of interest for specific ignitable liquid target compounds. The intensity profiles for selected ions are visually compared with profiles of known ignitable liquids. The mass-to-charge ratio of the ions of interest is selected using the typical ions observed from the table given by the ASTM method.[35] Unfortunately, interfering compounds may also be extracted when performing this manipulation.

Multivariate statistical techniques are useful in interpreting complex data. Principal compound analysis (PCA) and soft independent model classification analogy (SIMCA) are two data analysis techniques used in the multivariate analysis of ILRs. Principle component analysis is the linear transformation of an original set of measurements to a smaller set of uncorrelated values that retains information present in the original data set. SIMCA is a supervised learning technique used to create principle component analysis for each ignitable liquid class. While these techniques show some potential for improvements, there are still some questions before they can be applied routinely. Pattern recognition methods depend on relative abundances or resolved peaks to determine a match, and the presence of coeluting interfering compounds can affect these relative abundances.

8.6 Quality Assurance

One recent initiative in quality assurance has been the creation of the peer-based technical working group for fire and explosions (TWGFEX), initially sponsored by the National Institute of Justice (NIJ). The TWGFEX has created a scene response group whose mission it is to:

> ... establish and maintain nationally accepted programs for the forensic investigation of fire, arson, and explosion scenes and devices. Further, to promote and maintain dialogue among personnel in the public safety and legal communities.[75]

The TWGFEX laboratory analysis group has also published their mission statement as:

> The mission of TWGFEX-Lab is to make recommendations for nationally accepted guidelines for the forensic examination of fire and explosive materials and residues.[75]

The goals of the TWGFEX laboratory analysis group include performing collaborative exercises; specifying educational requirements for analysts' knowledge, skills, and abilities; establishing quality assurance guidelines; and striving to gain national acceptance of TWGFEX guidelines. The TWGFEX Web site can be reached at http://ncfs.ucf.edu/twgfex/home.html.

8.7 Conclusions

Many recent advances in the analysis of ILRs from fire debris have resulted from a significant amount of research activity in this area. New developments that improve the sensitivity and selectivity of the analysis by improving the extraction of the analytes hold promise, including the application of SPME for the sampling and extraction of ILRs. There is particular hope for field applications of SPME for ILR analysis. GC/MS/MS has been shown to offer improvements in both sensitivity and selectivity for the analysis of many components of ILR while allowing for the identification of target compounds in the presence of a background with large amounts of pyrolysis products. FT ICR/MS is a developing technique that holds promise for some specialized cases of ILR analysis. The future of ILR analysis is promising as these newly reported techniques become validated and gain wider adoption amongst the operational forensic laboratories.

References

1. J.R. Almirall and K.G. Furton, in Proceedings of the International Symposium on the Forensic Aspects of Arson Investigations, U.S. Government Printing Office, Washington, D.C., 1995, 337.

2. K.G. Furton, J.R. Almirall, and J. Bruna, *J. High Resolut. Chromatogr.*, 18, 625, 1995.

3. J.R. Almirall, J. Bruna, and K.G. Furton, *Sci. Justice; J. Forensic Sci. Society*, 36(4), 283, 1996.

4. K.G. Furton, J.R. Almirall, and J. Bruna, *J. Forensic Sci.*, 41, 12, 1996.

5. J.R. Almirall and K.G. Furton, in *Solid Phase Microextraction: A Practical Guide*, S.A. Scheppers Wercinski, Ed., Marcel Dekker, New York, 1999, 203.

6. K.G. Furton, J. Wang, Y.-L. Hsu, J. Walton, and J.R. Almirall, *J. Chromatogr. Sci.*, 38, 297, 2000.

7. J.R. Almirall, J. Wang, K. Lothridge, and K.G. Furton, *J. Forensic Sci.*, 45(2), 461, 2000.

8. K.G. Furton, J.R. Almirall, M. Bi, J. Wang, and L. Wu, *J. Chromatogr.*, 885, 419, 2000.

9. J.R. Almirall and K.G. Furton, in *Sample Preparation in Field and Laboratory*, J.B. Pawliszyn, Ed., Elsevier Science, Amsterdam, 2002, 919.

10. Q.L. Ren and W. Bertsch, *J. Forensic Sci.*, 44, 504, 1999.

11. A.C. Harris and J.F. Wheeler, *J. Forensic Sci.*, 48, 41, 2003.

12. J.R. Almirall and K.G. Furton, *J. Anal. Appl. Pyrolysis*, 2003, first published online on July 9, 2003.

13. R.W. Clodfelter and E.E. Hueske, *J. Forensic Sci.*, 22(1), 116–118, 1977.

14. P. Rapley, presented at a joint meeting of the Royal Society of Chemistry and the Association of Consulting Chemists and the Forensic Science Society, London, England, February 14, 1987; abstract: *J. Forensic Sci. Soc.*, 27(3), 210, 1987.

15. J.D. DeHaan and K. Bonaris, *J. Forensic Sci. Soc.*, 28(5), 299–309, 1988.

16. W. Bertsch, *J. Chromatogr.*, 674, 329–333, 1994.

17. W. Bertsch, *Forensic Sci. Rev.*, 9(1), 1–22, 1997.

18. J.R. Almirall and J. Perr, The use of compound specific MS/MS for the identification of ignitable liquid residues in fire debris analysis, in *Advances in Forensic Applications of Mass Spectrometry*, J. Yinon, Ed., CRC Press, Boca Raton, FL, 2003, in press.

19. D.A. Sutherland, *Can. Soc. Forensic Sci. J.*, 30, 185, 1997.

20. M.D. Plasencia, S. Montero, J. Krivis, A. Armstrong, and J. Almirall, Proceedings of the American Academy of Forensic Sciences Meeting, 2000.

21. J.M. Perr, C. Diaz, K.G. Furton, and J.R. Almirall, Proceedings of the International Association of Forensic Science Meeting, Montpellier, France, 2002.

22. B.J. deVos, M. Froneman, E. Rohwer, and D.A. Sutherland, *J. Forensic Sci.*, 47, 736, 2002.

23. J.M. Perr, K.G. Furton, and J.R. Almirall, Proceedings the American Academy of Forensic Sciences Meeting, Chicago, 2003.

24. J.R. Almirall, Mass Spectrometry in Forensic Science, Proceedings of the International Mass Spectrometry Conference, Edinburgh, U.K., 2003, in press.

25. Technical Working on Fire and Explosions Web site, University of Central Florida, http://ncfs.ucf.edu/twgfex/mission.html, July 30, 2003.

26. Fire and Arson Scene Evidence: A Guide for Public Safety Personnel, National Institute of Justice, http://www.ojp.usdoj.gov/nij/pubs-sum/181584.htm, June 2000.

27. Guide for Fire and Explosion Investigations, National Fire Protection Association (NFPA) 921, February, 2001.

28. ASTM E1412-00 Standard Practice for Separation of Ignitable Liquid Residues from Fire Debris Samples by Passive Headspace Concentration With Activated Charcoal, *ASTM Annual Book of Standards*, Vol. 14.02, ASTM International, West Conshohocken, PA, 2002.

29. ASTM E1413-00 Standard Practice for Separation and Concentration of Ignitable Liquid Residues from Fire Debris Samples by Dynamic Headspace Concentration, *ASTM Annual Book of Standards*, Vol. 14.02, ASTM International, West Conshohocken, PA, 2002.

30. ASTM E1388-00 Standard Practice for Sampling of Headspace Vapors from Fire Debris Samples, *ASTM Annual Book of Standards*, Vol. 14.02, ASTM International, West Conshohocken, PA, 2002.

31. ASTM E1385-00 Standard Practice for Separation and Concentration of Ignitable Liquid Residues from Fire Debris Samples by Steam Distillation, *ASTM Annual Book of Standards*, Vol. 14.02, ASTM International, West Conshohocken, PA, 2002.

32. ASTM E1386-00 Standard Practice for Separation and Concentration of Ignitable Liquid Residues from Fire Debris Samples by Solvent Extraction, *ASTM Annual Book of Standards*, Vol. 14.02, ASTM International, West Conshohocken, PA, 2002.

33. ASTM E2154-01 Standard Practice for Separation and Concentration of Ignitable Liquid Residues from Fire Debris Samples by Passive Headspace Concentration with Solid Phase Microextraction (SPME), *ASTM Annual Book of Standards*, Vol. 14.02, ASTM International, West Conshohocken, PA, 2002.

34. ASTM E1387-01 Standard Test Method for Ignitable Liquid Residues in Extracts from Fire Debris Samples by Gas Chromatography, *ASTM Annual Book of Standards*, Vol. 14.02, ASTM International, West Conshohocken, PA, 2002.

35. ASTM E1618-01 Standard Test Method for Ignitable Liquid Residues in Extracts from Fire Debris Samples by Gas Chromatography/Mass Spectrometry, *ASTM Annual Book of Standards*, Vol. 14.02, ASTM International, West Conshohocken, PA, 2002.

36. D.C. Mann and W.R. Gresham, *J. Forensic Sci.*, 35(4), 913–923, 1990.

37. K.P. Kirkbride, S.M. Yap, S. Andrews, P.E. Pigou, G. Klass, A.C. Dinan, and F.L. Pedie, *J. Forensic Sci.*, 37, 1585–1599, 1992.

38. J.R. Almirall, K.G Furton, and J.C. Bruna, Proceedings of the Southern Association of Forensic Scientists (SAFS) Fall Meeting, September 7–10, 1994, Orlando, FL.

39. T. Kaneko and M. Nakada, National Institute of Police Science Proceedings.

40. J. Perr and J.R. Almirall, Proceedings of the American Academy of Forensic Sciences 54th Annual Meeting, February 11–16, 2002, Atlanta, GA.

41. J. Perr, J.R. Almirall, and C. Diaz, Proceedings of the International Association of Forensic Scientists 16th Tri-Annual Meeting, September, 2002, Montpellier, France.

42. J. Handley and C.M. Harris, Great ideas of a decade, *Anal. Chem.*, 73, 23A–26A, 2001.

43. J. Pawliszyn, *Solid Phase Microextraction Theory and Practice*, Wiley-VCH, 1997.

44. W.E. Brewer, C.G. Randolph, S.L Morgan, and K.H. Habben, *J. Anal. Toxicol.*, 21, 286–290, 1997.

45. J. Pawliszyn, Unified theory of extraction, in *Sampling and Sample Preparation for Field and Laboratory*, J. Pawliszyn, Ed., 2002, 253–278.

46. http://www.gerstel.com/en/en_mainframe.html, December 12, 2002.

47. J.A. Lloyd and P.L. Edmiston, *J. Forensic Sci.*, 48, 130, 2003.

48. A. Stefeen and J. Pawliszyn, *Anal. Commun.*, 33, 129, 1996.

49. J. McCurry, *Organic Chemistry*, 4th ed., Brooks/Cole Publishing, New York, 1996.

50. D. Drysdale, *An Introduction to Fire Dynamics*, John Wiley & Sons, Chichester, 1985.

51. C.L. Thomas, Arson debris control samples, *Fire Arson Invest.* 28, 23, 1978.

52. D.J. Tranthim-Fryer and J.D. DeHaan, *Sci. Justice*, 37, 39, 1997.

53. W. Bertsch, Volatiles from carpet: A source of frequent misinterpretation in arson analysis, *J. Chromatogr. A*, 674, 329–333, 1994.

54. E. Stauffer, Identification and characterization of interfering products in fire debris analysis, thesis, Florida International University, Miami, 2001.

55. J.J. Lentini, Persistence of floor coating solvents, *J. Forensic Sci.*, 46, 1470–1473, 2001.

56. W. Bertsch, Chemical analysis of fire debris: Was it arson?, *Anal. Chem.*, 68, 541A–545A, 1996.

57. R. Newman, M. Gilbert, and K. Lothridge, *GC/MS Guide to Ignitable Liquids*, CRC Press, Boca Raton, FL, 1998.

58. Z. Liu and J.B. Philips, Comprehensive Two-Dimensional Gas Chromatography Using an On-Column Thermal Desorption Modulation Interface, *J. Chromatogr. Sci.*, 29, 227–231, 1991.

59. C.J. Venkatramani and J.B. Philips, Comprehensive two-dimensional gas chromatography (GC × GC) applied to the analysis of complex mixtures, *J. Microcol.*, 5, 511–516, 1995.

60. J. Blomberg, P.J. Shoenmakers, J. Beens, and R. Tijssen, Comprehensive two-dimensional gas chromatography (GC × GC) and its applicability to the characterization of complex (petrochemical) mixtures, *J. High Resolut. Chromatogr.*, 20, 539–544, 1997.

61. J. Beens, H. Boelens, R. Tijssen, and J. Blomberg, Quantitative aspects of comprehensive two-dimensional gas chromatography (GC × GC), *J. High Resolut. Chromatogr.*, 21, 47–54, 1998.

62. J. Beens, J. Blomberg, and P.J. Schoenmakers, Proper tuning of comprehensive two-dimensional gas chromatography (GC × GC) to optimize the separation of complex oil fraction, *J. High Resolut. Chromatogr.*, 23, 182–188, 2000.

63. G.S. Frysinger and R.B. Gaines, Forensic analysis of ignitable liquids in fire debris by comprehensive two-dimensional gas chromatography, *J. Forensic Sci.*, 47, 471–482, 2002.

64. R.B. Gaines, G.S. Frysinger, M.S. Hendrick-Smith, and J.D. Stuart, Oil spill source identification by comprehensive two-dimensional gas chromatography, *Environ. Sci. Technol.*, 33, 2106–2112, 1999.

65. L.M. Blumberg, Potentials and limits of comprehensive GC × GC, *Anal. Chem.*, 503A, 2002.

66. W. Paul and H. Steinwedel, U.S. Patent 2,939,952 (1960).

67. Z. Ziegler, Ion traps come of age, *Anal. Chem.*, 489A–492A, 2002.

68. J.F. Todd, Introduction to practical aspects of ion trap mass spectrometry, in *Practical Aspects of Ion Trap Mass Spectrometry*, R.E. March and J.F. Todd, Eds., CRC Press, Boca Raton, FL, 1995.

69. J.A. Hipple, H. Sommer, and H.A. Thomas, *Phys. Rev.*, 76, 1877–1878, 1949.

70. H. Sommer, H.A. Thomas, and J.A. Hipple, *Phys. Rev.*, 82, 697–702, 1951.

71. A.G. Marshall, C.L. Hendrickson, and S.D.H. Shi, *Anal. Chem.*, 253A–259A, 2002.

72. R.P. Rodgers, E.N. Blumer, M.A. Freitas, and A.G. Marshall, *J. Forensic Sci.*, 46, 268–279, 2001.

73. M.W. Gilbert, *J. Forensic Sci.*, 43, 871–876, 1998.

74. B. Tan, J.K. Harsy, and R.E. Snavely, *Anal. Chim. Acta*, 442, 37–46, 2000.

75. Technical Working Group for Fire and Explosions Web site, July 30, 2003, http://ncfs.ucf.edu/twgfex/home.html

Index